Effective Carbon Rates

PRICING CO_2 THROUGH TAXES AND EMISSIONS TRADING SYSTEMS

OECD
BETTER POLICIES FOR BETTER LIVES

Please cite this publication as:
OECD (2016), *Effective Carbon Rates: Pricing CO_2 through Taxes and Emissions Trading Systems*, OECD Publishing, Paris.
http://dx.doi.org/10.1787/9789264260115-en

ISBN 978-92-64-26009-2 (print)
ISBN 978-92-64-26011-5 (PDF)

Photo credits: Cover © west cowboy/Shutterstock.com, ©Nivellen77/Shutterstock.com

Corrigenda to OECD publications may be found on line at: *www.oecd.org/about/publishing/corrigenda.htm*.

Foreword

Climate change is one of the defining challenges of our age. Without drastic cuts in greenhouse gas emissions, the world faces devastating impacts. If policy responses are slow, these impacts are likely to be more costly than any economic crisis the world has experienced so far. Delaying the low carbon transition will exacerbate the costs, since taking action only when it is truly unavoidable will render carbon-intensive capital obsolete.

In response to this vital challenge, at the COP21 Conference on Climate Change in 2015, 195 countries agreed to decarbonise the global economy by the second half of the century. This requires deep change in the structure of modern economies to put them on a pathway to green growth.

Carbon pricing provides countries with a low cost tool to effectively and gradually reduce emissions starting immediately. While action on climate change needs to go beyond carbon pricing, it is an essential part of the solution. As this study makes clear, even with a broad definition of carbon pricing, 90% of carbon emissions from energy use today are priced below EUR 30 per tonne, which represents a conservative minimum estimate of the damage that results from emitting one tonne of carbon. There can be no question that delivering on the promises of COP21 requires strengthening carbon pricing now.

The *Effective Carbon Rates* report is unique in its comprehensive approach, integrating carbon prices that result from taxes and emissions trading systems. Taxes include carbon taxes and – importantly – specific taxes on energy use more generally. The analysis covers 41 OECD and G20 economies, responsible for 80% of global energy use and CO_2 emissions, and is broken down into six economic sectors for each country.

Most countries significantly price emissions from the road sector, while carbon emissions from the non-road sectors go largely unpriced: 70% of emissions are priced at zero and only 4% of emissions are priced above EUR 30 per tonne. Sectors outside road transport comprise primarily emissions from industry, electricity generation and the residential and commercial sectors, and they account for 85% of carbon emissions from energy use in the 41 countries. Higher carbon prices in these sectors will be an essential ingredient to an effective policy response.

Carbon pricing can be strengthened through both taxes and emissions trading. Emissions trading can be an effective option, but sustained investment in low and zero-carbon technologies requires higher and more stable permit prices than generally observed today. Taxes can be grafted onto existing tax systems and this simplifies implementation. Irrespective of the tools used, durable progress requires increasing carbon prices in most economies and in many sectors.

This report, by providing comprehensive and comparable information of the current state of CO_2 emissions pricing, can support policymakers to define priorities and improve carbon pricing policies. Even modest action taken by all countries can translate into strong progress. For example, if all countries would at least emulate carbon pricing structures of the median country in each economic sector, the extent to which emissions are priced at less than EUR 30 per tonne would decline from 80% at present to around 50%.

We simply cannot afford, environmentally or economically, to get these policies wrong. The carbon clock is ticking – our world needs higher effective carbon rates now, before time runs out.

Angel Gurría

Secretary-General, OECD

Acknowledgements

This report was prepared by Johanna Arlinghaus, Luisa Dressler, Florens Flues, Michelle Harding and Kurt Van Dender under the guidance of Kurt Van Dender. All authors are in the OECD's Centre for Tax Policy and Administration (CTPA), with Johanna Arlinghaus co-funded by the Environment Directorate for part of her involvement.

The authors would like to thank the following OECD colleagues for their very insightful feedback on earlier versions of the report and for their contributions to its development: Nils-Axel Braathen, David Bradbury, Olivier Durand-Lasserve, Nathalie Girouard, Takayoshi Kato, Youngnoh Kim, Aidan Kennedy (IEA), Martine Monza, Violet Sochay and Carrie Tyler.

Several experts from government, research institutions and international organisations have provided essential inputs to the estimation of the coverage of emissions trading systems. We mention in particular Brieanne Aguila and Anny Huang of the California Air Resources Board, United States; Jifeng Li of the State Information Center, China; Jan Christoph Nill of the European Commission; Sunhwa Chung of the Korean delegation to the OECD; Petra Icha of the Umweltbundesamt, Germany; Yasuaki Kaway of the Saitama Prefectural Government and Masahiro Kimura of the Tokyo Metropolitan Government, Japan; Nora Vogel of RGGI Inc; Reto Schaefer of the Federal Office for the Environment, Switzerland; and Steve Doucet-Héon and Vicky Leblond from the Ministère du Développement durable, de l'Environnement et de la Lutte contre les changements climatiques, Québec, Canada.

The report was discussed at various stages of its development by the OECD's Joint Meetings of Tax and Environment Experts, and it was approved for publication by the Committee of Fiscal Affairs and the Environment Policy Committee. Invaluable suggestions, inputs to and comments on the report were provided by the delegates to the Joint Meetings and their colleagues in national and sub-national government administrations. This project has been made possible with resources provided by the OECD Secretary-General's Allocation Fund.

Table of contents

Figures

Tables

Box

Abbreviations

ARB	Air Resources Board
COP	Conference of the Parties
CO_2	Carbon Dioxide
ECR	Effective Carbon Rate
EU	European Union
EIA	Energy Information Agency
ETS	Emissions Trading System
GDP	Gross Domestic Product
ICAP	International Carbon Action Partnership
IEA	International Energy Agency
INDC	Intended Nationally Determined Contribution
IMF	International Monetary Fund
NACE	Statistical classification of economic activities in the European Community
NAICS	North American Industry Classification System
OECD	Organisation for Economic Co-operation and Development
RGGI	Regional Greenhouse Gas Initiative
TEU	Taxing Energy Use
TMG	Tokyo Metropolitan Government
UNFCCC	United Nations Framework Convention on Climate Change
WHO	World Health Organisation

Executive summary

Tackling climate change requires deep cuts in greenhouse gas emissions, including CO_2 emissions. Carbon pricing is an effective and low cost means of inducing carbon abatement. While not in itself sufficient to deliver the degree of abatement required for limiting the risks of climate change, carbon pricing is an essential part of the solution. However, new evidence presented in this report shows that 90% of carbon emissions are not priced at a level reflecting even a conservative estimate of their climate cost.

This report presents the first comprehensive analysis of the extent to which countries use carbon prices. It measures effective carbon rates, the price of carbon emissions resulting from taxes and emissions trading systems, in 41 OECD and G20 countries, accounting for 80% of global energy use and of CO_2 emissions. For each country, the analysis shows the distribution of effective carbon rates across all energy use and their composition by six economic sectors.

Carbon prices often are zero or very low. Across the 41 countries, 60% of carbon emissions from energy use are unpriced. Where carbon is priced, the price tends to be low. The damage from climate change resulting from a tonne of CO_2 emissions can very conservatively be estimated at EUR 30. Only 10% of emissions are priced at an effective carbon rate equal or exceeding EUR 30 per tonne of CO_2. In other words, 90% of emissions fail the weak test of being priced at a level in accordance with a low end estimate of the climate damage they cause.

Across all countries, effective carbon rates are particularly low in sectors outside road transport, with 70% of emissions not priced at all and only 4% of emissions subject to an effective carbon rate above EUR 30. These sectors, which include the industry, electricity and commercial & residential sectors as well as offroad transport and agriculture & fisheries, emit 85% of carbon emissions from energy use in the group of 41 countries.

Specific taxes on energy use form the main component of effective carbon rates in the non-road sectors, with 23% of emissions from energy use subject to such a tax, and with specific tax rates on average higher than carbon tax or permit price levels. Emissions trading systems have a significant impact on coverage, and they extend the share of emissions subject to a price to 30%.

Road transport has comparatively high effective carbon rates, with 46% of emissions priced above EUR 30 per tonne of CO_2 and only 2% of emissions unpriced. These rates are almost entirely the result of specific taxes on road transport fuel, which are not usually introduced primarily for climate reasons. Fuel taxes can reflect air pollution and to some extent congestion costs, justifying rates well above EUR 30 per tonne of CO_2.

Within countries, effective carbon rates on transport fuels are higher than those on other energy use. In the non-road sectors, some countries mostly price emissions from industry, though often at low rates, while other countries price emissions from the commercial and residential sector more strongly. Emissions from the electricity sector are priced in many countries, mostly at low rates and often through consumption taxes, which do not encourage switching to cleaner fuels. The impact of emissions trading systems on effective carbon rates is largest in the industry and the electricity sectors. Taxes apply to 17% of emissions in industry and 27% in electricity, and price signals from emissions trading increase these shares to 26% for industry and 36% for electricity.

The report introduces the "carbon pricing gap" as a synthetic indicator of the extent to which effective carbon rates fall short of pricing emissions at EUR 30 per tonne of CO_2. If all emissions were priced at least at EUR 30, the carbon pricing gap would be zero, and if all emissions were unpriced, it would be 100%. Currently, the carbon pricing gap is 80.1% for the group of 41 countries. If carbon prices and coverage were to increase to at least the levels currently observed at the median for each economic sector, the carbon pricing gap would decline to 53.1%. This suggests that meaningful progress with carbon pricing can be made by increasing rates were they are currently low and introducing pricing instruments where they are zero. Such an approach would contribute to more uniform carbon prices, which improves cost-effectiveness.

The non-road sectors, and in particular the industry, the electricity and the commercial and residential sectors, are prime targets for raising rates. Within these sectors, non-oil based energy tends to face lower rates than oil products. Efforts could also be concentrated in countries where effective carbon rates are low across all energy use, although in some cases this may lead to calls for inter-country transfers. Such a gradual approach would lead to more uniform carbon prices, broadening the scope for cost-effective abatement. Cost-effectiveness is always desirable and becomes indispensable as abatement targets become stringent.

Progress with carbon pricing can be made by utilising taxes or emissions trading systems. Carbon pricing mechanisms co-exist with other mitigation policies. Ideally pricing should be the primary policy tool used to drive abatement, for if regulatory or other policies lead, cost-effectiveness likely suffers. Trading may have the advantage of better political feasibility, particularly when combined with free allocation of allowances. However, making durable progress with meaningful carbon pricing will require higher and more stable rates than currently observed in trading systems, as well as increased attention to revenue raising and productive ways of using revenue from taxes or auctioned emission permits. Turning towards taxes may be the simplest approach administratively, as carbon taxes can often be grafted onto existing systems, and the most effective approach economically, as creating well-functioning markets is not always straightforward.

Reader's guide

This report is divided into two main parts: Part I sets the scene and provides a cross-country analysis of effective carbon rates, Part II presents detailed information on the effective carbon rates for the 41 countries included in this report. Annexes A and B have additional detail on the methodology and data.

Part I is further divided into five different chapters:

Chapter 1 provides a short introduction into effective carbon rates and summarises the main results.

Chapter 2 is a recap of the policy and economic background to carbon pricing. Readers familiar with the arguments for the environmental- and the cost-effectiveness of carbon pricing and its wider economic effects can skip this chapter.

Chapter 3 defines effective carbon rates and details the concept and methodology underlying the results presented in this report. Readers wishing to focus purely on results can turn directly to Chapter 4.

Chapter 4 presents the results of the analysis of effective carbon rates (ECRs):

The first section of Chapter 4 ("Patterns of CO_2 emissions from energy use in 41 countries") sketches the patterns of CO_2 emissions from energy use across sectors and countries, providing context for the interpretation of the effective carbon rates.

The second section of Chapter 4 ("Effective carbon rates: Results of the analysis") presents the detailed results of the analysis of effective carbon rates:

- It starts with an overview of the distribution of ECRs across the 41 economies included in this report, both at an aggregate and a country level, separately for emissions from road and non-road sectors.
- This is followed by analysis of the composition of ECRs, i.e. the relative impacts of specific taxes on energy use, carbon taxes and emissions trading systems.
- The next subsection focuses on the level and composition of ECRs within the five non-road sectors, and specifically on ECRs on energy use in the industry sector and the electricity sector.
- The final subsection provides a brief description of how the analysis of ECRs changes when treating biomass as carbon neutral on a lifecycle basis instead of including emissions from the combustion of biomass in total emissions.

The third section of Chapter 4 ("Effective carbon rates: the bigger picture") places the effective carbon rates into a broader context:

- It starts with constructing an indicator, "the carbon pricing gap", to summarise the extent to which emissions are not priced at the reference level of EUR 30 used in this report.

- This is followed by a simulation of policy-induced changes of the carbon pricing gap indicator, assuming that all countries catch up with median country performance in all sectors.
- The final subsection relates ECRs and the share of emissions priced at different rates to some broader macroeconomic and energy characteristics of countries.

Chapter 5 sums up and concludes.

Part II presents details of the effective carbon rates in individual countries.

Chapter 6 shows detailed information on the distribution of effective carbon rates, and decomposes the effective carbon rates by price instrument for each of the 41 countries separately, both on an economy-wide level and by different sectors of the economy.

Annex A describes the detailed methodology and the data sources used for calculating ECRs.

Annex B provides additional information on estimating the coverage of and determining the permit prices for the sixteen emissions trading systems which are included in the analysis of effective carbon rates.

Part I

Effective carbon rates in OECD and selected partner economies

Chapter 1

Effective carbon rates: An introduction and main results

This chapter provides the context and motivation for the analysis of effective carbon rates, which are the total price that applies to CO_2 emissions from energy use as a result of market-based instruments (carbon taxes, specific taxes on energy use and price signals from emissions trading systems). The chapter also gives a short overview of the main results for 41 countries which together account for 80% of global carbon emissions from energy use.

Containing the adverse impacts of climate change, whether certain or likely, and limiting the risk of catastrophic climate change, requires drastic reductions of greenhouse gas emissions. Energy use accounted for 69% of total anthropogenic greenhouse gas emissions in 2010 and emissions of carbon dioxide (CO_2) form the largest portion (IEA, 2014). The demand for energy is expected to increase strongly in the first part of the 21st century and beyond, due to economic and population growth in many major economies. Reducing CO_2 emissions from energy use is therefore a critical part of any policy package to mitigate climate change.

The Paris Agreement adopted at the 21st Conference of the Parties to the United Nations Framework Convention on Climate Change (COP21) in December 2015 is a milestone in international efforts to craft an effective response to climate change. One of the aims of the Paris Agreement is to hold the global average temperature increase to below 2°C and to pursue efforts to limit it to 1.5°C above pre-industrial levels. Limiting temperature increases to around 2°C in 2100 will require greenhouse gas emissions to fall by between 40% and 70% of current levels by 2050, and they should be close to zero by 2100 (Evans, 2014). Achieving the more ambitious 1.5°C target even calls for reducing CO_2 emissions to zero in net terms between 2045 and 2060 (Rogelj et al., 2015).

Mitigation pledges submitted by individual countries in the lead up to COP21 are not sufficient to limit temperature increases to below 2°C, but instead are more likely to put the world on a pathway leading to a global temperature increase of about 3°C (Gütschow et al., 2015; Kitous and Keramidas, 2015). The Paris Agreement provides countries with a platform to share long-term goals, including a collective and cyclical review of nationally determined contributions to these goals every five years. If countries want to reach the agreed limit in global temperature increase, they will need to substantially strengthen their efforts to reduce emissions.

Carbon pricing is an effective and cost-effective means of reducing CO_2 emissions. This report provides the first comprehensive, detailed and systematic evidence on "effective carbon rates" in 41 countries covering 80% of global energy use. Effective carbon rates are the sum of carbon taxes, specific taxes on energy use and tradable emission permit prices, expressed in EUR per tonne of CO_2 emissions. They are, in other words, the price on carbon emissions that energy users face as a result of market-based policies. By providing this measure, the report is part of OECD work to support countries in strengthening their emission reduction efforts, including in the context of the Paris Agreement.

The principal appeal of using prices to induce carbon abatement is that this encourages emission reductions where they are cheapest, both in the sense of using the cheapest available options today and steering innovation and investment towards lower-carbon technologies. This is the main reason why, for example, the OECD, the World Bank, the IMF, the Inter American Center of Tax Administrations, the European Commission and the Inter-American Development Bank "share the view that while pricing CO_2 emissions alone may not be sufficient to address climate change, it is an essential part of the solution" (International Tax Dialogue, 2015). Implementing carbon pricing is not always straightforward, for example because of concerns about adverse impacts on equity and on competitiveness. The FASTER principles of carbon pricing provide insight on how progress can be made (OECD and World Bank Group, 2015).

The contribution of this report is threefold. First, the effective carbon rates (ECRs) provide a comprehensive view of the use of market-based policies, integrating prices resulting from carbon taxes, specific taxes on energy use, and tradable emission permit prices, into a comprehensive measure of carbon pricing in 41 OECD and G20 economies. Second, the full distribution of rates across all energy use is studied, allowing an appreciation of the considerable heterogeneity of carbon prices, e.g. by considering the share of emissions priced at various levels. Third, effective carbon rates are disaggregated by six economic sectors, showing in detail where CO_2 emissions are covered by a carbon price and from which type of instrument this price derives.

Available reports on carbon pricing consider either explicit pricing mechanisms (carbon taxes and tradable emissions permits) or taxes (carbon taxes and specific taxes on energy use) but have not considered all three market-based policies together. This report complements three sets of reports in particular. First, *Taxing Energy Use: A Graphical Analysis* (OECD, 2013) and *Taxing Energy Use 2015: OECD and Selected Partner Economies* (OECD, 2015a) provide a systematic analysis of the structure and level of carbon taxes and specific taxes on energy use by fuel type and sector in 41 OECD and G20 economies. As will be seen, specific taxes on energy use are currently the main component of effective carbon rates. The present report builds on the *Taxing Energy Use* work by integrating price signals from emissions trading systems. Second, *State and Trends of Carbon Pricing*, published by the World Bank Group (Kossoy et al., 2015), provides a global overview of all existing carbon taxes and emissions trading systems, but does not include specific energy taxes other than carbon taxes. Third, the *International Carbon Action Partnership's Status Report* (ICAP, 2016) gives detailed information on emissions trading systems in force and under consideration worldwide, including estimates of the share of emissions that are under the cap of each ETS.

Some of the main features of the effective carbon rates in the 41 countries are discussed in *Effective Carbon Rates on Energy – OECD and Selected Partner Economies* (OECD, 2015b), a brochure released during COP21. The present report provides much more detail, disaggregating the effective carbon rates by country and by six sectors within each. In addition, it analyses the structure of carbon pricing across and within the 41 countries. Technical detail on the methodology is also included.

One significant insight from the analysis is that carbon pricing is anything but universal. Across the 41 countries, the effective carbon rate is zero for 60% of all CO_2 emissions from energy use. Only 10% of emissions are priced at or above EUR 30 per tonne, a low-end estimate of the climate cost of emissions. Effective carbon rates are comparatively high in road transport, where 46% of emissions are subject to a rate of at least EUR 30 per tonne. In other sectors, 70% of emissions face an effective carbon rate of zero, and just 4% are priced at EUR 30 or more.

Although inter-country differences in levels of effective carbon rates are large, there are common intra-country patterns. The low prices across the group of countries as a whole are partly the consequence of low prices and limited coverage in several large countries. Six countries feature much lower rates on road transport emissions than the 35 other countries, although in all countries road transport faces much higher ECRs than emissions in other sectors. Outside road transport, only one country charges more than EUR 30 per tonne on more than half of its emissions from energy use. ECRs tend to be higher in countries where per capita GDP is higher and in those countries that import a larger share of energy. Higher ECRs also correlate with lower carbon-intensity of GDP.

Specific taxes on energy use, which are predominantly excise taxes, dominate the other two components of effective carbon rates (carbon taxes and tradable permit prices). Carbon taxes are low on average and cover only a small part of emissions from energy use across the group of countries considered. Tradable permit prices are also low, but contribute significantly to coverage of non-road emissions with a price. Just under a quarter (23%) of emissions from industry and electricity are subject to carbon taxes or specific taxes on energy use; when emissions trading systems are also taken into account, 30% of emissions in these sectors are subject to a price.

The "carbon pricing gap" is introduced as a synthetic indicator – measured as a percentage – of the extent to which effective carbon rates fall short of pricing emissions at EUR 30 per tonne, the low-end estimate of the cost of carbon that is used as a reference in this report. The carbon pricing gap is 80.1% across the 41 countries as a whole. A counterfactual scenario is calculated in which carbon rates and their coverage are increased to the level of the median country rate and coverage in the six economic sectors distinguished in the analysis (leaving rates and coverage at their current levels if they already exceed these of the median country). This reduces the carbon pricing gap to 53.1%.

The counterfactual scenario suggests that meaningful progress with carbon pricing is within reach, not least by introducing prices on emissions currently not priced or by increasing the price where effective carbon rates are presently low. Prime targets for a gradual approach to increasing coverage and rates include emissions from non-road transport sources and from fuels other than oil. Efforts could also be concentrated in countries where prices presently are low across the board, although in some cases this may lead to calls for inter-country transfers. The gradual approach contributes to more uniform effective carbon rates, and so broadens the scope for cost-effective emission reductions.

Progress with carbon pricing can be made by utilising a range of instruments. Emissions trading presently covers a portion of emissions from energy use (to some extent, but not fully, overlapping with the coverage of taxes) but at low rates. Trading may have the advantage of political feasibility, particularly when combined with free allocation of allowances. However, making durable progress with meaningful carbon pricing will require higher and more stable rates than currently observed in trading systems, as well as increased attention for revenue raising and for productive ways of using revenue from taxes or auctioned emission permits.

Carbon pricing mechanisms co-exist with other mitigation policies. Ideally, market-based instruments should be the primary policy tool used to drive abatement, for if regulatory or other policies lead, cost-effectiveness likely suffers. With regard to the type of carbon pricing, turning towards taxes may be the simplest approach administratively (as carbon taxes can often be grafted onto existing systems) and the most effective one economically (as creating well-functioning markets is not always straightforward), particularly if they are sufficiently uniform and political commitment over time is strong. Interactions between taxes and emissions trading systems also need to be kept in mind. Taxing emissions subject to a trading system does not result in additional emission reductions as long as the cap is binding, and may even be counterproductive if it destroys uniformity of carbon prices produced by the trading system and the cost-effectiveness associated with such uniform prices. This does not necessarily mean that taxes and trading systems should never be combined, but the justification for levying taxes on ETS-covered emissions should be to address other market failures or to raise revenue.

Countries may of course opt for less cost-effective carbon abatement policies, for example when market-based policies are politically very difficult to implement whereas other effective policies are within reach. These alternative policies also put implicit prices on carbon, as they induce costly abatement. However, lower carbon-intensities of GDP are observed in countries with higher effective carbon rates. To the extent that the causality runs from higher rates to lower carbon-intensity, this suggests that the implicit carbon prices from non-market based instruments, where they exist, have not triggered the same level of abatement as the effective carbon rates have elsewhere. If further abatement is sought, as climate targets strongly suggest it should, then aiming for least cost strategies becomes a more important consideration, and the appeal of market-based instruments once again increases.

References

Evans, S. (2014), "Two degree: Will we avoid dangerous climate change", Carbon Brief, www.carbonbrief.org/two-degrees-will-we-avoid-dangerous-climate-change (accessed 26 Jan 2016).

Gütschow, J. et al. (2015), "INDCs lower projected warming to 2.7˚C: significant progress but still above 2°C", Climate Action Tracker Update, http://climateactiontracker.org/news/224/indcs-lower-projected-warming-to-2.7c-significant-progress-but-still-above-2c-.html (accessed 26 Jan 2016).

ICAP (2016), *Emissions trading Worldwide: Status Report 2016*, International Carbon Action Partnership, Berlin, https://icapcarbonaction.com/images/StatusReport2016/ICAP_Status_Report_2016_Online.pdf.

IEA (2014), *CO_2 emissions from fuel combustion – Highlights – 2014 Edition*, OECD/IEA, Paris.

International Tax Dialogue (2015), "Much better use can and must be made of taxes to help reduce pollution and greenhouse gas emissions", www.oecd.org/environment/much-better-use-can-and-must-be-made-of-taxes-to-help-reduce-pollution-and-greenhouse-gas-emissions.htm (accessed 26 Jan 2016).

Kitous, A. and K. Keramidas (2015), "Analysis of scenarios integrating the INDCs", *European Commission Joint Research Centre Policy Brief*, European Commission, Joint Research Centre, Seville.

Kossoy, A. et al. (2015), *State and trends of carbon pricing 2015*, World Bank Group, Washington, DC.

OECD (2015a), *Taxing Energy Use 2015: OECD and Selected Partner Economies*, OECD Publishing, Paris, http://dx.doi.org/10.1787/9789264232334-en.

OECD (2015b), *Effective Carbon Rates on Energy: OECD and Selected Partner Economies*, OECD, Paris, https://www.oecd.org/tax/tax-policy/effective-carbon-rates-on-energy.pdf.

OECD (2013), *Taxing Energy Use: A Graphical Analysis*, OECD Publishing, Paris, http://dx.doi.org/10.1787/9789264183933-en.

OECD and World Bank Group (2015), *The FASTER principles for successful carbon pricing: An approach based on initial experience*, OECD, Paris and World Bank Group, Washington, DC, https://www.oecd.org/environment/tools-evaluation/FASTER-carbon-pricing.pdf (accessed 26 Jan 2016).

Rogelj, J. et al. (2015), "Energy system transformation for limiting end-of-century warming to below 1.5 °C", *Nature Climate Change*, Vol. 5, pp. 519-528.

Chapter 2

Carbon pricing: Reducing emissions in a cost-effective manner

This chapter discusses why carbon prices are an effective and low-cost policy tool to reduce emissions. It starts by discussing the environmental effectiveness of carbon prices. This is followed by a brief consideration of why carbon prices allow countries to reach their emissions targets in the cheapest possible way, while at the same time implementing the polluter pays principle and boosting economic benefits. The chapter also discusses climate cost of carbon emissions, and considers carbon pricing in a broader economic context.

This chapter explains why carbon pricing is an attractive means of addressing climate change. The first section ("Why carbon prices are effective") discusses environmental effectiveness, i.e. why carbon prices effectively reduce emissions. The second section ("Why carbon prices are cost-effective") considers cost-effectiveness, i.e. the ability of carbon pricing to reduce emissions at the lowest cost to society. The third section ("Carbon prices help implement the polluter pays principle and boost economic benefits") explains how pricing emissions adds economic benefits by obtaining a better balance between more and less emission-intensive activities. The fourth section ("The climate cost of carbon") discusses different estimates of the climate cost of carbon emissions. Lastly, the fifth section ("Carbon pricing in a wider economic context") considers the impacts of carbon pricing on innovation, on competitiveness and on equity among households. Carbon prices induce innovation and can have a positive effect on the level of productivity in a country. Concerns about potential adverse effects of carbon prices on industrial competitiveness and on equity often do not materialise and can, if needed, be addressed by appropriate policy design.

Why carbon prices are effective

Carbon prices are effective for reducing emissions because they increase the price of carbon-based energy, so decreasing demand for it (cf. Arlinghaus, 2015; Martin et al., 2013). Pricing carbon can lead to substitution towards less carbon-intensive forms of energy and to lower demand for energy overall. Taking generation of electricity as an example, producers can switch from coal to natural gas, or to zero-carbon energy sources, such as solar and wind power. In addition, where market structure and regulation allow, electricity producers will pass on the increase in production costs resulting from carbon prices to consumers of electricity, in the form of higher electricity prices, and this encourages price-sensitive consumers to decrease consumption. For example, firms and households may become more vigilant about turning off appliances when they are not in use or may use them less, and they can choose more efficient appliances at the time of replacement.

In general, the extent to which carbon prices reduce emissions depends on the level of the carbon price and on the availability of substitutes for the emitting activity. Larger price increases usually generate larger emissions reductions, and where substitutes are expensive or not available, emissions abatement may occur only at higher carbon price levels. Where substitutes are readily available at small additional cost, emission abatement can be substantial even at relatively low carbon price levels. The scope for substitution varies between sectors and between countries.

Empirical analysis finds that price elasticities of energy demand differ from zero, and there is strong evidence that long-run elasticities are larger than short-run elasticities. Estimates of long-run price elasticities differ among studies, depending on type of energy and user, country, and time period; methodological choices and data types also affect estimates. To give an order of magnitude, estimates of (absolute) values of long-run own price elasticities of demand often fall in the 0.3 to 0.8 range (see, e.g. EIA, 2014; Graham and Glaister, 2002; Labandeira et al., 2012); this holds both for household and industrial demand. The meta-analysis by Labandeira et al. (2015) finds a long-term own-price elasticity of -0.6 across various fuels, with small differences across fuels. This means that a 10% price increase would lead to a reduction in demand by 6%. Taking a bird's eye and very long-run view, Fouquet (2014) finds that (absolute values of) price elasticities of household energy use tend to be higher at lower and higher levels of income than at middle income levels, and that the demand for domestic heating is less responsive to price increases than the demand for transport and lighting.

Empirical work on price elasticities of energy demand tends to focus on own-price elasticities of end-user demand. Cross-price elasticities and the input fuel demand of electricity producers are less studied. A carbon price will encourage electricity suppliers and end-users to switch from dirtier to cleaner energy sources. Estimated own-price elasticities of energy demand do not inform directly on these margins, and in that sense provide a lower bound for emission reductions to be expected from carbon prices.

Aside from the availability of energy substitutes, the environmental effectiveness of carbon prices depends on how well other policies are aligned with carbon pricing and emissions reduction efforts. Alignment requires that barriers (e.g. regulations or market structure) do not prevent price signals from triggering action. OECD (2015) identifies substantial misalignments between climate policy and regulatory frameworks across a range of policy domains. One example relates to the regulation of electricity markets, which in many countries imposes market designs that inadvertently are more favourable towards fossil fuel power generation than to the use of renewables. A different example relates to the preferential income tax treatment of company cars found in many OECD countries. Such treatment may reduce effective labour income tax rates, but often leads to more and bigger cars that are used more intensively, with considerable impacts on emissions (Harding, 2014; Roy, 2014). As a last example, the effectiveness of carbon pricing depends on billing structures for energy products too. If, for example, households pay for heating according to the size of their apartment, but not according to the amount of heating energy they actually use, incentives to conserve heating energy are very weak.

Why carbon prices are cost-effective

Carbon prices are a cost-effective policy tool to reduce emissions, and this makes them particularly attractive compared to other policy options. Cost-effectiveness means that an abatement target is reached at minimum cost. The cost-effectiveness of carbon pricing is due to several factors. Firstly, only emitters that can cut emissions more cheaply than paying the price will do so, while emitters facing higher abatement costs can continue to emit and pay the price. Secondly, emitters can choose to reduce emissions in the least costly way to themselves. Finally, cost-effectiveness has an important dynamic aspect, in that all emitters have an ongoing incentive to reduce their carbon bill by finding new and cheaper ways to cut emissions (OECD, 2010).

Carbon prices decentralise abatement decisions, which overcomes the asymmetry of information between the government that wishes to abate emissions, and the emitters that have to take abatement action. Firms and households are in the best position to determine their response to carbon prices, as they have the best information on the available options. Taking private car use as an example, households can respond to an increase in energy prices by driving less, driving more economically, purchasing a more fuel-efficient car, or switching to cheaper fuel types. While a particular response may be suitable for one household, it will turn out to be less so for another, and households themselves are best informed on what works for them and what does not.

Policies that constrain emitters' choices, in contrast, risk imposing high adaptation costs. For example, a fuel economy standard, which regulates the average distance travelled per unit of fuel for cars, may force low-mileage households into investing in fuel efficiency with low or even negative private payoffs, and only limited reductions in aggregate fuel use. Such a standard may not reduce driving but can instead increase it, as the fuel cost per kilometre declines (Small and Van Dender, 2007; Frondel et al., 2012). Bans on driving on specific days or in specific places also impose high costs on households and businesses, since they do not distinguish between low- and high-value trips.

Cost-effectiveness is not just a theoretical property. Comparing different emissions reduction policies in 15 countries, OECD (2013a) finds that a carbon price is the policy tool with the lowest cost of emission reductions. Learning effects (not considered in OECD, 2013) can reduce the cost of low-carbon technologies and thereby cut the cost of future emission reductions, and this can justify the use or improve the relative performance of non-pricing instruments, but it does not follow that these instruments obviate the need for carbon pricing, as the subsidies needed to reduce costs of low-carbon technologies are lower when carbon prices are in place.

Ensuring that a uniform carbon price covers as many emissions as possible ("base broadening"), maximises cost-effectiveness. The more emissions a carbon price covers as a share of a country or a region's total emissions, the more scope exists for cost-effective mitigation, hence the lower the carbon price needed to reduce total emissions by the aspired amount. Setting a uniform carbon price across all sectors of an economy allows abatement at lowest cost, which is tantamount to saying that such an approach permits the lowest carbon price needed for a given abatement target. Sectors with sufficiently cheap mitigation opportunities will abate more than sectors where the options are more costly. For example, a carbon price of EUR 30 per tonne of CO_2 may encourage an industrial bakery to invest gradually in more energy-efficient ovens, but it is unlikely to make the bakery replace its truck fleet for deliveries. At the same time, a price of EUR 30 per tonne of CO_2 will likely induce the bakery's electricity supplier to change its electricity production from coal to a cleaner source.

Carbon prices help implement the polluter pays principle and boost economic benefits

Carbon prices improve resource allocation within an economy by ensuring that firms, private and public sector organisations, and households pay for the damage resulting from their CO_2 emissions (the polluter-pays-principle). Without a carbon price, energy users will equate their marginal private benefits and costs from energy use. Many harmful side-effects of energy use, such as climate change or air pollution, which affect society more broadly and not only the energy user responsible for the emissions, are not normally taken into account in private consumption decisions, i.e. they are external costs. Carbon prices that are aligned with the social cost of CO_2 emissions make energy users incorporate the harmful climate effects of energy use into their consumption decisions, i.e. they internalise the external costs of climate change. This leads to more efficient use of resources, which is equivalent to saying that the net benefits from all economic activity rise. Net benefits rise because the external costs of climate change have been reduced from their previously excessive levels, given that emitters now have to pay for emissions.

The European Systemic Risk Board (2016) outlines the likely consequences of implementing serious climate policy only at the very last minute. Late and sudden climate policy implementation will likely come at a higher economic cost compared to a smooth introduction of carbon pricing and complementary climate policies, for three reasons. First, abrupt implementation of hard constraints on emissions from fossil fuels will severely affect the production of goods and services because of more limited access to alternative fuels in this case. Second, markets may suddenly re-price carbon-intensive assets that are largely financed by debt, and this can impact financial stability. Third, extreme weather events may affect financial stability via their impacts on the liabilities of insurers and reinsurers.

The climate cost of carbon emissions

To ensure carbon prices are effective in causing emitters to take the damage of their emissions into account, their levels should be aligned with the marginal cost of climate change from each tonne of CO_2 emissions. Alternatively, prices can be set as a function of policy targets on abatement, which may lead to prices higher or lower than estimated damage. This report focuses on the first approach, which is referred to as the climate cost of CO_2 emissions. Estimating this climate cost is difficult given uncertainties over the climatic and economic processes involved, and the long term over which these processes will play out. As a consequence, available estimates differ and are uncertain. This report uses EUR 30 per tonne of CO_2 as a lower-end estimate of climate cost, which is in line with the values from a thorough review of recent evidence (Alberici et al., 2014). The same study arrives at a central value of EUR 50 per tonne of CO_2. Smith and Braathen (2015) find that the unweighted average cost of carbon used in policy appraisal in 2014 was slightly above USD 50 (approximately EUR 40).

The cited estimates relate to current levels. However, economic analysis, e.g. in integrated assessment models, prescribes price paths that imply significant increases of real carbon prices over time. For example, the shadow price of a tonne of CO_2 proposed in a French analysis starts at EUR 32 in 2010, rising to EUR 56 in 2020, EUR 100 in 2030 and EUR 200 in 2050 (Centre d'analyse stratégique, 2009). Similarly, to limit the temperature increase to 1.5 °C in line with the more ambitious target of the Paris Agreement, models indicate that a carbon price of EUR 50 per tonne of CO_2 is required today but that it needs to increase to hundreds of EUR per tonne of CO_2 in 2050 (Rogelj et al., 2015).

Comparing effective carbon rates to the lower bound of EUR 30 per tonne of CO_2 can be seen as a weak test, in the sense that rates should be at least at that level if they are to reflect climate costs. If, as it turns out is the case, by far the largest share of emissions is not priced at this level, this means that the use of carbon pricing fails to pass even a weak test. Selecting this lower-end estimate as a benchmark does not imply that carbon prices are sufficiently high at EUR 30 per tonne of CO_2, so this is not a policy conclusion that should be drawn from this report. Similarly, the use of EUR 30 does not represent a normative statement about the minimum level of pricing that should be implemented. Rather, it is a reference point which allows comparison of pricing policies across and within countries. Phasing in carbon prices over time can bring substantial benefits in practice. Starting with a relatively low price level but ensuring that the share of economy-wide emissions covered by a carbon price is large, broadens incentives to abate emissions. A low initial price can be a starting point for future price increases; such increases should be predictable if they are to send strong investment signals.

Extending carbon pricing to include more heating fuels and fuels used in industry and for electricity generation is likely to induce stronger emission reductions than increasing the carbon price on transport fuels where they are already high. This is not to say that there can be no case for increasing transport fuel prices (there often is), but in countries where they are already high, the climate-related justification for doing so may be weak. Unlocking the abatement potential in transport appears to require other policies in addition to carbon pricing.

Pricing CO_2 emissions can produce important co-benefits to CO_2 emissions mitigation. These co-benefits often are domestic, for example reductions in air pollution and other negative side-effects of energy use. Co-benefits can help make the domestic case for pricing energy use or emissions from energy use, and can lead to upward revisions of what price to strive for. Co-benefits are not quantified in the estimates of climate cost presented here, but the costs of, for example, air pollution are high (see OECD, 2014, and WHO and OECD, 2015), and their existence further supports the idea of charging at least the climate cost of CO_2 emissions from energy use. Of course, tackling non-climate external costs related to energy use should not run entirely via carbon prices, as different externalities require different policy approaches, including customised pricing policies.

Carbon pricing in a wider economic context

Environmental policy, with prices or other policy instruments, does not occur in an economic vacuum, and some interactions with the wider economy are worth considering. Here, revenue-raising effects, potential impacts on productivity growth, and competitiveness and distributional impacts are briefly discussed.

Carbon pricing can raise valuable public revenue, e.g. when emission permits are auctioned or taxes are levied. Judicious use of the revenues collected from newly introduced or increased carbon prices can generate additional economic benefits (OECD and World Bank, 2015). For example, using the revenues to lower capital or labour income tax rates results in efficiency gains if these income taxes are more harmful to economic activity than carbon prices. Whether this is the case or not is an empirical, context-dependent matter. Political economy and equity considerations can require other forms of revenue use, e.g. a degree of compensation for unduly adverse impacts on some firms or households, or agreements to use some revenues to stimulate research into green technologies or for climate finance. Grandfathering of permits can be seen as being similar to a tax expenditure, in the sense that potential revenues from auctioning are forgone. Such arrangements should be used sparingly and preferably be limited in time. Poor decisions on revenue use are likely to weaken the economic appeal of market-based instruments.

Environmental policies, including carbon pricing, can affect productivity and economic growth. A meta-analysis by Cohen and Tubb (2015) observes that econometric models which allow for the estimation of long-run effects show significant positive effects of environmental regulation on long-run productivity growth. Johnstone et al. (2010) find that the stimulating effect of environmental policies on technological progress is stronger for more stringent, more predictable and more flexible environmental policies. A higher, more predictable carbon price hence likely causes more innovation, and more so as larger portions of total emissions are covered.

Among the various instruments available to reduce CO_2 emissions, carbon prices are the most likely to foster short-term economic growth and increase the level of productivity in a country (Albrizio et al., 2014a). A potential positive effect of tighter environmental regulation on productivity is also found to be more robust for market-based instruments, such as carbon prices, than for other environmental policy instruments (ibid.). Cohen and Tubb (2015) confirm these findings, stating that flexible environmental policies are more likely to have a positive significant impact on productivity growth than less flexible ones. As argued above, carbon pricing offers more flexibility than other instruments, and the more broadly it is applied, the more scope for cost-reducing responses is created.

Adverse equity impacts of carbon pricing are a potential concern. The implementation or increase of prices on CO_2 emissions from energy use is often challenged by concerns that higher energy prices would hit poor or low-income households particularly hard. However, analysis produces a much more nuanced picture, showing that distributional impacts differ between fuels and countries, and that potential adverse equity effects can be addressed by accompanying policies. Flues and Thomas (2015) show for twenty-one, predominantly European, countries that the distributional impact of taxes on household energy use differs between fuels and countries. In many countries, taxes on transport fuels are not regressive, i.e. the burden as a share of household income or expenditure is not higher for low-income households.

Sterner (2012) finds that in low- and middle-income countries, transport fuel taxes are generally progressive, because high-income households spend a larger share of their income on transport fuels than low-income households. The share of low-income households owning a motor vehicle will be lower than among higher-income groups in these countries, so that taxes on transport fuels will not directly affect them, and any indirect effects through increased transport costs for other goods and services are likely modest. Taxes on heating fuels and electricity are often observed to be mildly regressive (Flues and Thomas, 2015), yet accompanying policies can counteract or reverse these impacts. For example, if carbon prices on energy use are to be increased, they generate revenue that can be distributed back to poor households in a targeted manner. Existing welfare systems may already counter, or can be adjusted to counter, increasing energy prices. Households can also be provided with targeted advice on how to become more energy-efficient (Tews, 2012).

Policy makers might fear reductions in industrial competitiveness as a result of carbon pricing policies. The econometric evidence reviewed in Arlinghaus (2015) and Partnership for Market Readiness (2015) shows that the potential adverse effects of carbon pricing on firm competitiveness have not materialised so far. Most reviewed studies find that carbon prices reduce emissions, but do not observe any meaningful competiveness effects. The results could potentially be explained by governments having successfully implemented exemptions or rebates from carbon prices for the firms expected to be most affected by carbon prices. However, studies that explicitly compare firms that face different carbon prices, but are otherwise similar, also do not find any consistent negative competitiveness impacts (Flues and Lutz, 2015; Gerster, 2015; Martin et al., 2014). This suggests that current carbon price levels do not harm competitiveness, even for those firms not granted preferential rates.

The carbon prices analysed in the studies on competitiveness are not very high, effectively between about EUR 5 to EUR 45 per tonne of CO_2. If carbon prices increase substantially in the future, it cannot be excluded that this may have impacts on competitiveness. However, if more countries implement a price on carbon, then competitiveness effects become less likely. In addition, the available evidence strongly suggests that gradually introducing carbon prices does not pose competitiveness threats to industries, even if individual firms can be hurt. This means that gradual progress with carbon pricing is feasible without triggering competitiveness alerts.

Albrizio et al. (2014b) observe that while the most productive firms tend to benefit from carbon pricing, less productive firms may find it more challenging to stay in business when carbon prices rise. In this situation, providing relief becomes a trade-off between helping the least productive firms, which may struggle for reasons other than carbon pricing, or to use carbon pricing revenue for other, potentially more productive and inclusive goals. Examples include supporting workers to better align their skills with the needs of a less manufacturing-based, more knowledge-based economy or using the revenues from carbon pricing to reduce other taxes that are likely more harmful to economic activity.

In sum, carbon pricing is an effective policy, which reduces emissions at least cost and improves the resource allocation of an economy by making polluters pay for the climate cost of their emissions. Studies on the wider economic effects of environmental regulation emphasise that flexible policies, such as carbon prices covering CO_2 emissions from the broadest possible set of sources, likely have a stronger effect on innovation and productivity growth than less-flexible, more prescriptive policies.

References

Alberici, S. et al. (2014), "Final Report and Annex 3" in *Subsidies and costs of EU energy*, Ecofys by order of the European Commission, https://ec.europa.eu/energy/sites/ener/files/documents/ECOFYS%202014%20Subsidies%20and%20costs%20of%20EU%20energy_11_Nov.pdf.

Albrizio, S., et al. (2014a), "Do Environmental Policies Matter for Productivity Growth?: Insights from New Cross-Country Measures of Environmental Policies", *OECD Economics Department Working Papers*, No. 1176, OECD Publishing, Paris, http://dx.doi.org/10.1787/5jxrjncjrcxp-en.

Albrizio, S., T. Koźluk and V. Zipperer (2014b), "Empirical Evidence on the Effects of Environmental Policy Stringency on Productivity Growth", *OECD Economics Department Working Papers*, No. 1179, OECD Publishing, Paris, http://dx.doi.org/10.1787/5jxrjnb36b40-en.

Arlinghaus, J. (2015), "Impacts of Carbon Prices on Indicators of Competitiveness: A Review of Empirical Findings", *OECD Environment Working Papers,* No. 87, OECD Publishing, Paris, http://dx.doi.org/10.1787/5js37p21grzq-en.

Centre d'analyse stratégique (2009), "Le valeur tutélaire du carbone", Rapport de la commission présidée par Alain Quinet, Rapports et documents, No. 16-2009, République Française.

Cohen, M. A. and A. Tubb (2015), "The Impact of Environmental Regulation on Firm and Country Competitiveness: A Meta-Analysis of the Porter Hypothesis", available at SSRN: http://dx.doi.org/10.2139/ssrn.2692919.

EIA (2014), *Price elasticities for energy use in buildings of the United States*, Energy Information Administration, US Department of Energy, Washington, DC.

European Systemic Risk Board (2016), "Too late, too sudden: Transition to a low-carbon economy and systemic risk", *Advisory Scientific Committee Report*, No. 6, European Systemic Risk Board, Frankfurt.

Flues, F. and B. J. Lutz (2015), "Competitiveness Impacts of the German Electricity Tax", *OECD Environment Working Papers,* No. 88, OECD Publishing, Paris, http://dx.doi.org/10.1787/5js0752mkzmv-en.

Flues, F. and A. Thomas (2015), "The distributional effects of energy taxes", *OECD Taxation Working Papers,* No. 23, OECD Publishing, Paris. http://dx.doi.org/10.1787/5js1qwkqqrbv-en.

Fouquet, R. (2014), "Long run demand for energy services: Income and price elasticities over two hundred years", *Review of Environmental Economics and Policy*, Vol. 8, No. 2, pp. 186-207.

Frondel, M., N. Ritter and C. Vance (2012), "Heterogeneity in the rebound effect: Further evidence for Germany", *Energy Economics*, Vol. 34, No. 2, pp. 461-467.

Gerster, A. (2015), "Do Electricity Prices Matter? Plant-Level Evidence from German Manufacturing", Paper presented at the Meetings of the European Economic Association, 2015, Mannheim.

Graham, D. and S. Glaister (2002), "The demand for automobile fuel – A survey of elasticities", *Journal of Transport Economics and Policy*, Vol. 36, No. 1, pp. 1-26.

Harding, M. (2014), "Personal Tax Treatment of Company Cars and Commuting Expenses: Estimating the Fiscal and Environmental Costs", *OECD Taxation Working Papers*, No. 20, OECD Publishing, Paris, http://dx.doi.org/10.1787/5jz14cg1s7vl-en.

Johnstone, N., I. Hascic and M. Kalamova (2010), "Environmental policy characteristics and technological innovation", *Economia Politica*, Vol. XXVII, No. 2, pp. 275-299.

Labandeira, X., J.M. Labeaga and X. Lopéz-Otero (2012), "Estimation of elasticity price with incomplete information", *Energy Economics*, Vol. 34, No. 3, pp. 627-633.

Labandeira, X., J.M. Labeaga and X. Lopéz-Otero (2015), "A meta-analysis on the price elasticity of energy demand, *Economics for Energy Working Paper*, No. 04/2015.

Martin. R., L.B. de Preux and U.J. Wagner (2014), "The impact of a carbon tax on manufacturing: Evidence from microdata", *Journal of Public Economics*, Vol. 117, pp. 1-14.

Martin, R., M. Muûls and U. Wagner (2013), "Impact of the EU ETS on Regulated Firms: What is the Evidence after Eight Years?" Draft prepared for a symposium issue of the Review of Environmental Economics and Policy.

OECD (2015), *Aligning policies for a low carbon economy*, OECD Publishing, Paris, http://dx.doi.org/10.1787/9789264233294-en.

OECD (2014), *The costs of air pollution: Health impacts of road transport*, OECD Publishing, Paris, http://dx.doi.org/10.1787/9789264210448-en.

OECD (2013), *Effective Carbon Prices*, OECD Publishing, Paris, http://dx.doi.org/10.1787/9789264196964-en.

OECD (2010), *Taxation, Innovation and the Environment*, OECD Publishing, Paris, http://dx.doi.org/10.1787/9789264087637-en.

OECD and World Bank Group (2015), *The FASTER principles for successful carbon pricing: An approach based on initial experience*, OECD, Paris and World Bank Group, Washington, DC, https://www.oecd.org/environment/tools-evaluation/FASTER-carbon-pricing.pdf (accessed 26 Jan 2016).

Partnership for Market Readiness (2015), *Carbon Leakage: Theory, Evidence and Policy Design*, World Bank Group, Washington, DC.

Rogelj, J.et al. (2015), "Energy system transformation for limiting end-of-century warming to below 1.5 °C", *Nature Climate Change*, Vol. 5, pp. 519-528.

Roy, R. (2014), "Environmental and Related Social Costs of the Tax Treatment of Company Cars and Commuting Expenses", *OECD Environment Working Papers*, No. 70, OECD Publishing, Paris, http://dx.doi.org/10.1787/5jxwrr5163zp-en.

Small, K. A. and K. Van Dender (2007), "Fuel Efficiency and Motor Vehicle Travel: The Declining Rebound Effect", *The Energy Journal*, Vol. 28, No. 1, pp. 25–51.

Smith, S. and N.A. Braathen (2015), "Monetary Carbon Values in Policy Appraisal: An Overview of Current Practice and Key Issues", *OECD Environment Working Papers*, No. 92, OECD Publishing, Paris, http://dx.doi.org/10.1787/5jrs8st3ngvh-en.

Sterner, T. (2012), *Fuel taxes and the poor – The Distributional effects of gasoline taxation and their implications for climate policy*, RFF Press, New York.

Tews, K. (2012), "Einzelprojektevaluierung: Stromspar-Check in einkommensschwachen Haushalten" (Evaluation of the energy savings check in low-income households), in *Evaluierung des nationalen Teils der Klimaschutzinitiative des Bundesministeriums für Umwelt, Naturschutz und Reaktorsicherheit* (*Evaluation of the national climate action plan of the German Ministry for the Environment*), Öko-Institut, arepo consult, Freie Universität Berlin, ecologic, FiFo Köln and J. Ziesing, Berlin.

WHO Regional Office for Europe and OECD (2015), "Economic cost of the health impact of air pollution" in *Europe: Clean air, health and wealth*, WHO Regional Office for Europe, Copenhagen.

Chapter 3

Effective carbon rates: Concept and scope

This chapter explains the concept of effective carbon rates, and provides an overview of the methodology and data sources used for their estimation. The treatment of emissions from the combustion of biomass, tax expenditures, fossil fuel support and value added taxes is discussed.

This short chapter defines the concept of effective carbon rates (ECRs) and provides an overview of the data used for their estimation. It discusses the treatment of emissions from biomass in the calculations, and the interaction between ECRs and fossil fuel support. Technical detail on the methodology used to estimate ECRs can be found in the annexes to this report.

Effective carbon rates: Definition

Effective carbon rates are the total price that applies to CO_2 emissions from energy use as a result of market-based policy instruments. They are the sum of taxes and tradable emission permit prices, and have three components (Figure 3.1):

- carbon taxes, which typically set a tax rate on energy based on its carbon content,
- specific taxes on energy use (primarily excise taxes), which are typically set per physical unit or unit of energy, but which can be translated into effective tax rates on the carbon content of each form of energy, and
- the price of tradable emission permits, regardless of the permit allocation method, representing the opportunity cost of emitting an extra unit of CO_2.

Figure 3.1. **Components of effective carbon rates**

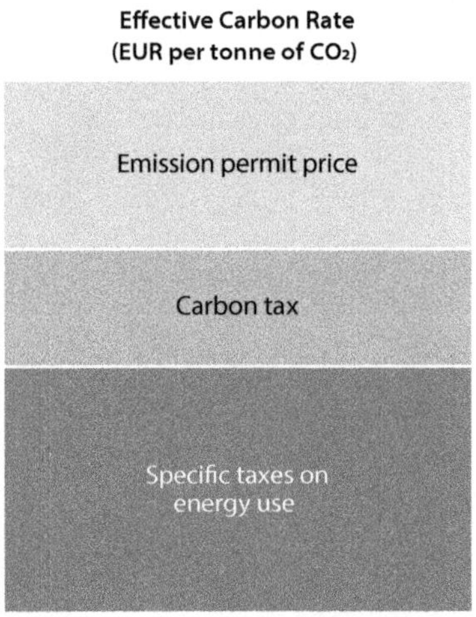

Source: OECD (2015c), *Effective Carbon Rates on Energy: OECD and selected partner economies*, https://www.oecd.org/tax/tax-policy/effective-carbon-rates-on-energy.pdf.

> Box 3.1. **Interpretation and cross-country comparisons of effective carbon rates**
>
> The effective carbon rates, summarised and discussed in Part I and depicted on a country level in the graphical profiles in Part II of this report, include specific taxes on energy use, carbon taxes and emission permit prices. In other words, they include those taxes and permit prices that alter the relative price of energy use, which is the relevant category of instruments from an environmental pricing point of view, based upon the principle that specific taxes and permit prices can alter relative prices to reflect marginal environmental damages.

Box 3.1. **Interpretation and cross-country comparisons of effective carbon rates** *(continued)*

Consistent application of this approach renders the effective carbon rate profiles amenable to inter-country comparison. Nevertheless, in some cases data limitations limit inter-country comparability. As an example of data limitations, the tax profiles do not account for differentiated VAT rates on energy products. Such differentiated rates alter relative prices and should therefore be accounted for in principle. The section on the "Treatment of tax expenditures, support for fossil fuels and value added taxes" in Chapter 3 discusses this issue in some detail. It is also worth noting that the report allows comparison of specific energy taxes, but not necessarily of total taxes on energy use. To illustrate, if country A and country B have the same excise tax on diesel for road use, but country A applies a lower standard VAT rate than country B, then the total tax rate on diesel for road use is lower in country A than in country B, but the specific rate is the same. This report focusses on the specific rate, for the reasons outlined previously in this box.

A further point is that the three components of the effective carbon rates are included irrespectively of how the revenues raised through these instruments are used. Specific taxes on energy use are included irrespectively of the policy rationale underlying their introduction. For example, specific taxes on energy use contribute to general revenue in their great majority, but in some countries, specific taxes on transport energy are used to fund infrastructure, whereas other countries use road tolls for the same purpose. The first are included in the effective carbon rates, the second are not, as the tax base for road tolls is not transport energy use.

For countries with a currency other than the Euro it is worth recalling ECRs and their components are converted into Euros using the average exchange rate over the 6 months prior to 1 April 2012. When a non-Euro currency appreciates, taxes converted into Euros rise although tax rates within the country do not change. If the appreciation were temporary, a non-Euro country will show higher tax rates in Euros than would be the case for a longer term average. Vice versa, a country with a temporary depreciation of its currency will show lower rates.

Data on the specific taxes on energy use and the carbon taxes are taken from the *Taxing Energy Use* database (TEU database, see also the following section for detail). Consistent with the *Taxing Energy Use* methodology, "the taxes levied on electricity consumption are included into the calculation of effective carbon rates on the fuels used to generate electricity. In cases where a common nominal tax rate is applied to all electricity consumption, the effective tax rate on each underlying energy source (e.g. coal, natural gas, hydro) used to generate electricity is shown. Notably, when there is a general tax on electricity consumption that applies regardless of the generation source, and if carbon energy is a small proportion of the generation mix, the effective tax rate on carbon thus calculated will be high. A tax on electricity consumption that does not distinguish between electricity from carbon sources does not send an effective price signal about the use of carbon" (OECD, 2013, 2015a).

Finally, it deserves emphasis that the report expresses the components of the effective carbon rates in EUR per tonne of CO_2, irrespective of the units in which the statutory rates of these components are set, and irrespective of the different external costs against which the different components could be compared. For example, transport fuel taxes can be seen as a second-best instrument to internalise a range of external costs of road use, including e.g. congestion, noise, and air pollution, in addition to climate damage from CO_2 emissions. This implies that ideally the prevailing taxes on transport energy should be compared to the full range of external costs that they are intended to cover. Such comparison is beyond the scope of the report, but it is nevertheless clear that an effective carbon rate of more than EUR 30 is justifiable. The first reason is that EUR 30 per tonne of CO_2 is a low end estimate of climate costs from carbon alone. The second reason is that other external costs may matter, and these can be high.

Box 3.1. **Interpretation and cross-country comparisons of effective carbon rates** *(continued)*

The message in this report is not that an effective carbon rate of EUR 30 per tonne of CO_2 is necessarily sufficient for internalising external costs. Instead, the message is that a rate below EUR 30 per tonne of CO_2 is too low to reflect conservative estimates of the external climate costs of energy use alone. Higher estimates of external climate costs and inclusion of other external costs can justify significantly higher rates. Major shares of energy use, however, fail to meet this modest level.

Data for estimating effective carbon rates

ECRs are estimated for 41 countries and six sectors: road transport, offroad transport, industry, agriculture and fisheries, residential and commercial and electricity. Tax rates and energy use data are for 2012. Some countries have introduced substantial tax reform since 2012. While such reforms can be mentioned in the country chapters, they are not included in the present discussion for reasons of methodological consistency. Price data for emissions trading systems are for 2012 or the year nearest to 2012 during which the system was operational and price data are available, as is explained briefly in this section and in more detail in Annex A.

Data on two components of the ECRs, namely the specific taxes on energy use and the carbon taxes, are taken from the *Taxing Energy Use* database (TEU database). The TEU database underlies the detailed analysis presented in *Taxing Energy Use: A Graphical Analysis* (OECD, 2013) and *Taxing Energy Use 2015: OECD and Selected Partner Economies* (OECD, 2015a), referred to together as the TEU reports. This database provides comprehensive information on the rates and coverage of carbon and other specific energy taxes in 41 countries. Taxes on energy use are translated from their original units in physical or energy terms to effective tax rates in terms of the carbon content of the fuels to which they apply. The present report adds prices of tradable emission permits, where they apply, to the information on tax rates from the TEU database. In addition, it estimates the share of emissions that are covered by emissions trading systems. The estimates of the coverage of emissions trading systems rely on a specific modelling approach and set of assumptions, a highly detailed discussion of which is in Annexes A and B.

Table 3.1 lists the emissions trading systems that are included in the calculation of ECRs. The systems operate in 30 of the 41 countries and cover approximately 13% of total CO_2 emissions from energy use in these countries.

To make data on emissions trading systems comparable across countries both the share of CO_2 emissions from energy use covered by emissions trading systems, and the emission permit prices, are estimated from detailed system data for 2012. Emissions trading systems that started operating at a later date are evaluated as if they were operational in 2012 and consumption price indices correct prices for inflation. First, emissions data to estimate the coverage of emissions trading systems are taken from verified emissions registries where available; where not, they are approximated by permit allocation data or information on fuel use. Second, emission permit prices are estimated by a yearly average, or as close an approximation as possible given data limitations or the duration of operation of the system. Annex A describes these estimations in detail, including assumptions specific to individual emissions trading systems.

Table 3.1. **Emissions trading systems operate in 30 countries**

Emissions trading system	Country	Introduced in
Beijing Emissions Trading System	China	2013
California Cap-and-Trade Program	United States	2013
Chongqing Emissions Trading System	China	2014
European Union Emissions Trading System	31 European countries; of which 23 OECD member countries	2005
Guangdong Emissions Trading System	China	2013
Hubei Emissions Trading System	China	2014
Korea Emissions Trading Scheme	Korea	2015
New Zealand Emissions Trading Scheme	New Zealand	2008
Québec Cap-and-Trade System	Canada	2013
Regional Greenhouse Gas Initiative	United States (9 north-east and mid-Atlantic US states)	2009
Saitama Prefecture Emissions Trading System	Japan	2011
Shanghai Emissions Trading System	China	2013
Shenzhen Emissions Trading System	China	2013
Swiss Emissions Trading Scheme	Switzerland	2008
Tianjin Emissions Trading System	China	2013
Tokyo Cap-and-Trade Programme	Japan	2010

An emissions trading system generally applies to only a portion of a country's emissions. Its coverage is calculated by country and sector as a share, dividing the sector emissions subject to the emissions trading system by the country's total emissions in that sector. This is done independently of whether the system applies in the entire country or to a subnational jurisdiction. In the case of a supranational ETS (i.e. the EU ETS), shares of coverage have been calculated for the six sectors separately for each country in which the system applies. For subnational trading systems, coverage is calculated as a proportion of national emissions. As a result, this report provides for the first time consistent coverage estimates of emissions trading systems for six economic sectors in the 41 countries.

To construct the effective carbon rate estimate, the coverage and price signals of emissions trading systems are combined with those of taxes from the *Taxing Energy Use* database, across the full range of fuel types and users (see Annex A for more information on how these datasets were combined). This allows for a systematic comparative analysis of the combined price signals that energy users face from both types of market-based mechanisms. In cases where taxes do not apply because a source is subject to an ETS, these taxes are not included in the tax rate shown in the figures.

Treatment of CO_2 emissions from the combustion of biomass

Emissions trading systems frequently treat CO_2 emissions from the combustion of biomass differently than CO_2 emissions from other sources, due to differences in approaches to accounting for their net lifecycle emissions. In several emissions trading systems, CO_2 emissions from the combustion of biomass are not included. In the remaining systems, emissions from biomass are subject to the ETS when emitted by a facility subject to the ETS, but are zero-rated when calculating covered emissions, meaning that no permit needs be surrendered for biomass emissions.

A separate point relates to the treatment of CO_2 emissions from biomass in the results. This report includes results for two different approaches: "Biomass included" calculates total emissions in each country including CO_2 emissions from biomass at the point of combustion, which is the approach taken in the TEU reports (OECD, 2013 and OECD, 2015a), and "Biomass excluded" treats emissions from biomass at combustion as carbon neutral and does not include them in the calculation of total emissions. The subsection on the "Treatment of biomass in the calculations" in Chapter 4 and Annex A provide more detail.

Treatment of tax expenditures, support for fossil fuels and value added taxes

The effective carbon rate analysis considers all emissions trading systems and taxes on energy use in each of the 41 countries covered, including energy which is not subject to either instrument. However, it does not take into account support measures for fossil fuel use that may affect its price. The OECD's database of budgetary support and tax expenditures, discussed in the *OECD Companion to the Inventory of Support Measures for Fossil Fuels 2015* (OECD, 2015b), identifies and documents almost 800 individual policies that support the production or consumption of fossil-fuels in OECD and six selected partner economies. The inventory includes direct budgetary transfers and tax expenditures that provide a benefit for fossil-fuel production or consumption when compared to alternatives. Many of these tax expenditures are included in the *Taxing Energy Use* analysis, and therefore are reflected in the effective tax rates used for the calculation of ECRs.

Including support measures for fossil fuels as measured in OECD (2015b) but not accounted for in the *Taxing Energy Use* work, could affect energy prices in ways not captured in the effective carbon rates. If support measures outweigh the tax and emissions trading components of end-user prices calculated in the present report, energy prices may even turn out to be negative. However, because of the complexity involved in combining taxes and emissions trading systems into effective carbon rates, this report makes no systematic attempt to incorporate support measures for fossil fuels, while noting that removing support for fossil fuels is a very cost-effective policy for reducing CO_2 emissions, and referring interested readers to OECD (2015c) for an in-depth discussion.

Value-added taxes (VAT) affect end-user prices of energy products in many jurisdictions in addition to the three components of ECRs. VAT is usually not specific to energy products: as long as the same VAT rate applies, the relative prices of energy products remain unchanged. Differential VAT rates, however, do change the relative prices of energy products and VAT then becomes a *de facto* specific tax measure. When this is the case, policy makers may want to consider the impact of the differential VAT rates in changing the relative prices of energy products in conjunction with the effective carbon rates. OECD (2015b) provides an overview of the differential VAT rates that apply in the 41 countries. Seventeen countries apply reduced or zero VAT rates on selected energy products. This counteracts the intention to increase the relative end-user prices of energy products, and can mitigate or even offset the effective carbon rate, depending on the relative magnitude of the amount of price differentiation introduced by the differential VAT rate and the effective carbon rate.

The caveats about support measures and the potential impact of VAT notwithstanding, the ECRs are an important component – and in most of the 41 countries analysed, by far the dominant component – of policy-driven carbon pricing.

References

OECD (2015a), *Taxing Energy Use 2015: OECD and Selected Partner Economies*, OECD Publishing, Paris, http://dx.doi.org/10.1787/9789264232334-en.

OECD (2015b), *OECD Companion to the Inventory of Support Measures for Fossil Fuels 2015*, OECD Publishing, Paris, http://dx.doi.org/10.1787/9789264239616-en.

OECD (2015c), *Effective Carbon Rates on Energy: OECD and Selected Partner Economies*, OECD, Paris, https://www.oecd.org/tax/tax-policy/effective-carbon-rates-on-energy.pdf.

OECD (2013), *Taxing Energy Use: A Graphical Analysis*, OECD Publishing, Paris, http://dx.doi.org/10.1787/9789264183933-en.

Chapter 4

Effective carbon rates: Results of the analysis

This chapter presents the results of the analysis of effective carbon rates. It first discusses the patterns of carbon emissions in the 41 countries included in the analysis. This is followed by an overview of the level of effective carbon rates in the road and non-road sectors, a presentation of the effective carbon rates by country and the composition of effective carbon rates by price instrument, as well as an analysis of effective carbon rates in the five non-road sectors (offroad, industry, agriculture and fisheries, residential and commercial, and electricity). The chapter also discusses the treatment of biomass in the calculations of effective carbon rates. This chapter introduces the "carbon pricing gap", which measures the extent to which emissions are priced at less than EUR 30 per tonne of CO_2, and uses this indicator to consider a counterfactual scenario of carbon pricing. The chapter closes by correlating effective carbon rates with countries' broader macroeconomic characteristics.

This chapter presents the results of the analysis of effective carbon rates (ECRs) on energy use in 41 OECD and selected partner economies. Together, these countries accounted for 80% of global CO_2 emissions from energy use in 2012. Across these countries, ECRs are found to be low, particularly outside road transport: 96% of emissions from energy use outside road transport are subject to an ECR of less than EUR 30 per tonne and 70% of non-road emissions are entirely unpriced.

To measure the degree to which emissions in the 41 economies are not priced, or are priced at less than EUR 30 per tonne of CO_2, a "carbon pricing gap" indicator is developed, expressed as a percentage of the situation where all emissions would be priced at EUR 30 per tonne. Under the policies analysed in this report, the carbon pricing gap presently amounts to 80.1%. However, if all countries priced carbon in every sector at least as much as the median country, the carbon pricing gap would be reduced to just above 53%.

This chapter begins with an analysis of the patterns of CO_2 emissions to be considered when interpreting ECRs in the first section ("Patterns of CO_2 emissions from energy use in 41 countries"). The second section ("Effective carbon rates: Results of the analysis") presents detailed results of the analysis of ECRs in the 41 countries, considering first the results for the group of countries as a whole and economy-wide results for each country, distinguishing between road and all other emissions, before discussing the role of each of the three instruments in the road and non-road sectors. The second section continues by examining the pattern of effective carbon rates in non-road sectors more closely and finishes by considering the treatment of biomass emissions for countries with larger shares of energy from biomass. The third section ("Effective carbon rates: The bigger picture") considers the broader context, introducing the "carbon pricing gap" indicator and using it to assess the effects of potential future carbon pricing policies. Finally, correlations between countries' effective carbon rates and the carbon intensity of GDP, GDP per capita and the share of net energy imports are explored.

Patterns of CO_2 emissions from energy use in 41 countries

The 41 countries for which this report presents ECRs together account for about 80% of global CO_2 emissions from energy use in 2012 (Figure 4.1). The countries are the 34 OECD countries and seven selected partner economies, namely Argentina, Brazil, China, India, Indonesia, Russia and South Africa, all of which are G20 countries. In 2012, the OECD share of CO_2 emissions from energy use was around 35% and that of the selected partner economies around 44%. The share of the selected partner economies is expected to grow rapidly, along with their share in world output. The OECD's share in global output is expected to decline from 62% in 2013 to 43% in 2050 (Johanssen et al., 2013).

Countries differ strongly in terms of energy use per capita and in the carbon-intensity of energy use (Figure 4.2). Many of the countries for which energy use per capita is set to grow strongly currently rely on relatively carbon-intensive energy sources. In the figure, the curved dotted lines denote three different levels of CO_2 emissions per person (5, 7.5 and 10 tonnes of CO_2 per year), for each level of carbon intensity of energy use (vertical axis) and the corresponding level of energy use per capita (horizontal axis). To reduce CO_2 emissions per capita, countries face the challenge of moving towards the lower left corner of Figure 4.2, where CO_2 emissions per capita are low. Cutting CO_2 emissions per capita is a matter of reducing energy use per person, i.e. moving "left" in Figure 4.2, or a matter of decarbonising energy use, i.e. moving "down", or both. The mix of both approaches will differ between countries.

Figure 4.1. **OECD and selected partner economies account for the bulk of CO_2 emissions from energy use**

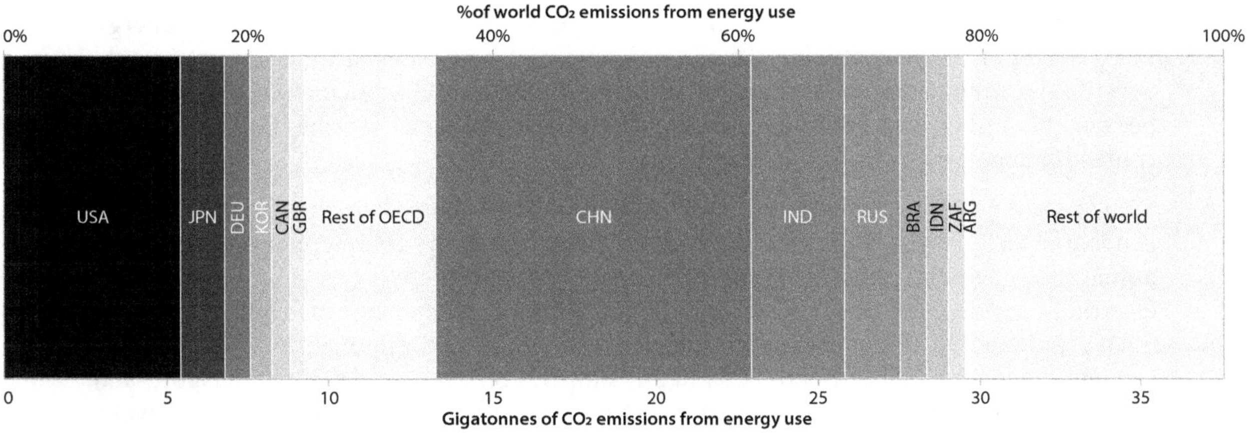

Note: CO_2 emissions data for 2012.

Source: OECD calculations based on OECD (2015), *Taxing Energy Use 2015: OECD and Selected Partner Economies*, http://dx.doi.org/10.1787/9789264232334-en, and IEA (2016), "Extended world energy balances", *IEA World Energy Statistics and Balances* (database), http://dx.doi.org/10.1787/data-00513-en.

Figure 4.2. **Energy use per capita and carbon intensity vary widely across countries**

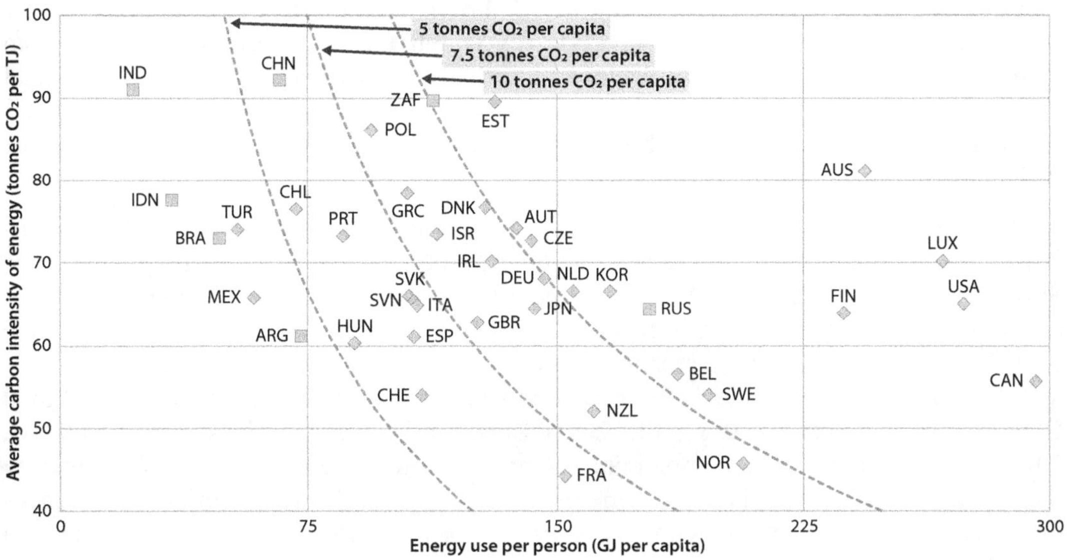

Note: The curved dotted lines show equal levels of CO_2 emissions per person and year, i.e. 5, 7.5 and 10 tonnes of CO_2 per capita. CO_2 emissions per person decrease when moving towards the lower left corner of the figure.

Source: Adapted from OECD (2015), *Taxing Energy Use 2015: OECD and Selected Partner Economies*, http://dx.doi.org/10.1787/9789264232334-en.

Prices can provide incentives for moving "down" and "left". A price proportional to the energy content of a fuel can reduce energy use (horizontal axis), also reducing carbon emissions. A price based on the carbon content of the fuel discourages the use of carbon-based fuels and particularly carbon-intensive fuels. These prices can both decrease the carbon intensity of the energy mix (vertical axis) and reduce energy use (horizontal axis).

For example, a price based on the carbon content of a unit of energy from bituminous coal would be around 1.75 times as high as a price on the carbon content on a unit of energy from natural gas. A price based on energy content would be the same per unit of energy independent of the fuel. Pricing the carbon content of both fuels at the same rate would constitute a stronger signal to move to natural gas than would a price on energy content, because bituminous coal would be priced at a higher rate under a price based on carbon content than on a price based on energy content.

Figure 4.3 shows that the proportion of CO_2 emissions from energy use in each economic sector (road transport, offroad transport, industry, commercial & residential, agriculture & fishing, and electricity) varies widely across countries. The industry and electricity sectors each account for roughly a third of carbon emissions from energy use (33%), followed by the road sector, and the residential and commercial sector, which account for roughly 15% of emissions in all countries on average. Offroad transport, and agriculture and fishing emit 2% and 1% of carbon emissions from energy use, respectively.

In road transport, the proportion of total emissions ranges from just 6% in China to 69% in Luxembourg. A general pattern is that CO_2 emissions from energy use in road transport tend to be higher in countries with a higher GDP per capita. A few small landlocked countries (Luxembourg, Switzerland and Slovenia) have a particularly high share of carbon emissions from road transport. In some of these countries, this is because of high shares of transit traffic in total road transport and because of non-residents filling up their vehicles, especially when transport fuels are priced at a lower level than abroad. Higher prices for transport fuels will likely lead to lower recorded emissions in these countries, but would largely be compensated by increases in nearby countries. The share of emissions from road transport in total emissions in Iceland is particularly high because of the high share of energy use generated from renewables causing no carbon emissions. Due to their particular characteristics, Iceland and Luxembourg are treated as outliers when citing ranges for the emissions from road transport.

Carbon emissions from energy use in industry range from about 10% of total emissions in Israel, Luxembourg and Greece to about 60% of total emissions from energy use in the Slovak Republic, Finland and Sweden (excluding biomass from the calculation reduces this to 51% in Finland and 45% in Sweden). This range reflects country differences in the contribution of the industry sector to economic output, but also differences in the fuels used by industry. For example, in Luxembourg the industry sector accounts for 11% of total GDP while it accounts for 30% in the Slovak Republic (own calculations based on OECD, 2016). The residential and commercial sector contributes between 1% of total energy related emissions in Iceland and 39% in Switzerland. Again, this range reflects differences in the share of the residential and commercial sector in overall economic activity as well as differences in the fuels used by commerce and households. While pricing carbon or increasing the level or coverage of carbon pricing instruments may be unlikely in the short to medium term to alter markedly the weight of the industry or commercial and residential sector in the overall economy of the 41 countries, energy users in these sectors will be encouraged to lower emissions by switching to cleaner fuels and reducing energy use.

Carbon emissions from energy used in electricity generation range from close to zero in Iceland, Switzerland, Norway and Luxembourg to 40% of domestic emissions and well above in Korea, South Africa, Australia, Greece and Israel. With the exception of Luxembourg, which imports a large share of its electricity use from neighbouring countries, all other countries with a share of carbon emissions from electricity use at close to zero generate their electricity mostly from carbon-free, mostly renewable, energy sources. In contrast, the countries with high shares of carbon emissions from energy use in electricity generation rely heavily on fossil fuels, in particular on coal. In these countries, pricing carbon can be expected to encourage electricity generators to switch from coal to cleaner fuels and thereby lower emissions.

In most countries, the offroad transport sector and the agriculture and fisheries sector contribute substantially less to total carbon emissions from energy use than the previously

Figure 4.3. **The composition of CO_2 emissions from energy use by sector varies widely across countries**

Note: Figure 4.3 sorts countries by their share of CO_2 emissions from the road transport sector in 2012.

Source: OECD calculations based on OECD (2015), *Taxing Energy Use 2015: OECD and Selected Partner Economies*, http://dx.doi.org/10.1787/9789264232334-en, and IEA (2016), "Extended world energy balances", *IEA World Energy Statistics and Balances* (database), http://dx.doi.org/10.1787/data-00513-en.

discussed sectors (c.f. Table 4.3 in the subsection on "Effective carbon rates within the non-road sectors"). Nevertheless, carbon emissions from offroad transport are a more substantial share of the total in some countries, with e.g. 9% of total emissions in Norway and about 5% in the United States, Canada and Russia. Whereas CO_2 emissions from energy use in pipelines are mainly responsible for this large share in Russia, domestic aviation and navigation are the main contributors in the United States and Norway respectively. With the exception of Iceland, emissions from agriculture and fisheries are at about 5% of total emissions or below. While emissions from these sectors tend to be relatively low, some of the sectors, such as air transport, may in the future contribute a substantially larger share, especially if no measures are undertaken to limit their emissions.

Effective carbon rates: Results of the analysis

A complete picture of how countries use market-based instruments to price CO_2 emissions requires evaluating all three components of the effective carbon rate (ECR) together, i.e. carbon taxes, specific taxes on energy use and tradable emission permit prices. Also, when evaluating ECRs, it is crucial to consider both the rate and the share of emissions facing a price (the latter being called ECR coverage in the remainder). This section presents detailed results of the analysis of ECRs both across countries as a whole and for individual countries, and discusses how the individual instruments contribute to the overall level and coverage of carbon prices.

Effective carbon rates – overview and country results

This subsection provides an overview of the ECRs for the group of countries as a whole and of ECRs at the economy-wide level for each individual country. Due to the differences between road and other sectors, ECR levels and coverage are presented separately for CO_2 emissions from road transport (road) and for those from all sectors other than road transport (all non-road sectors), including offroad transport, industry, agriculture and fishing, residential and commercial, and electricity.

ECRs across OECD and selected partner economies

Across the 41 OECD and selected partner economies, 60% of all CO_2 emissions from energy use do not face a carbon price from market-based instruments: the ECR is zero. Where CO_2 emissions are priced, the price is generally very low: 90% of all emissions face an ECR below EUR 30 per tonne of CO_2, the lower-end estimate of the climate cost of carbon used here. Prices on emissions from sectors other than road transport – "non-road sectors", responsible for 85% of total CO_2 emissions in the 41 countries – are particularly low. In road transport, nearly all emissions have a positive and relatively high ECR because of high specific taxes on energy use in this sector. Specific taxes on energy use provide the main price signal, both in terms of the level of ECRs and in terms of the proportion of emissions facing an ECR larger than zero, even outside the road sector. Carbon taxes and tradable emission permit prices currently constitute only a small share of ECRs. These patterns hold both across the 41 economies and within most individual countries.

Figure 4.4 summarises the ECRs that apply to all CO_2 emissions from energy use across the 41 countries. The information is presented by ECR intervals, showing the share of emissions that are not priced, and the shares of emissions with an ECR between EUR 0 and EUR 5, between EUR 5 and EUR 30, and above EUR 30 per tonne of CO_2. The left panel relates to all CO_2 emissions from energy use in the 41 countries, the middle panel shows emissions in the road transport sector, and the right panel shows emissions from all non-road sectors.

Figure 4.4. **Proportion of CO_2 emissions from energy use at different ECR intervals**

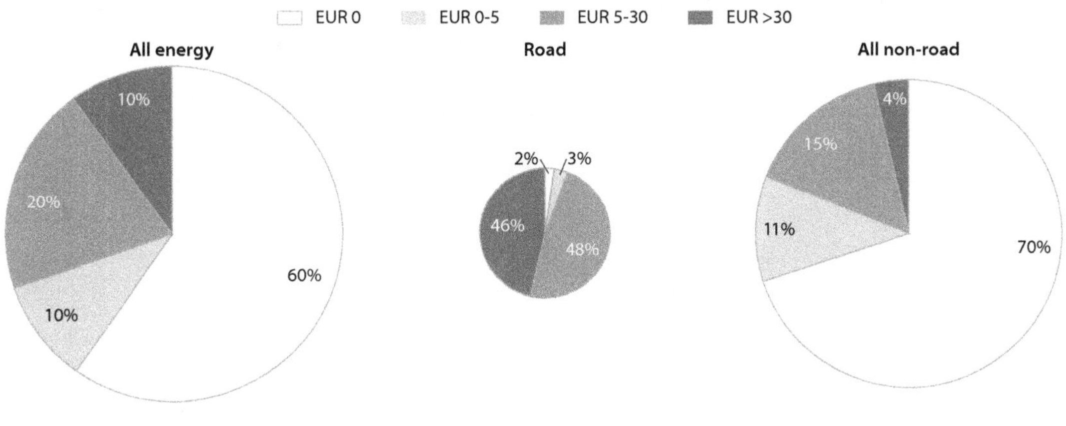

Note: The size of each pie chart corresponds to the amount of CO_2 emissions in the respective sector: all energy (100%), road transport (15%), and all non-road sectors (85%).

Considering total energy use in the 41 countries (left panel of Figure 4.4), 60% of CO_2 emissions are not priced at all, 10% are subject to an ECR between EUR 0 and EUR 5 per tonne of CO_2, 20% to an ECR between EUR 5 and EUR 30 per tonne of CO_2, and 10% to an ECR above EUR 30 per tonne of CO_2. Hence, 90% of total CO_2 emissions from energy use in the 41 countries are priced below the lower-end estimate of the climate cost of carbon. 70% of total CO_2 emissions are priced at a level of less than EUR 5 per tonne of CO_2, implying that there is only a very small policy-driven price incentive to reduce CO_2 emissions.

Economy-wide ECRs hide considerable differences between road and non-road sectors. In road transport (middle panel of Figure 4.4), only 2% of CO_2 emissions are not priced at all and 46% face an ECR above EUR 30 per tonne of CO_2. Excluding emissions from road transport (right panel of Figure 4.4), 70% of CO_2 emissions are not priced at all and only 4% face an ECR above EUR 30 per tonne of CO_2. The relatively high ECR in road transport is a result of specific taxes on energy use (i.e. excise taxes on gasoline and diesel) that are applied at comparatively high rates in each of the 41 countries – with few exceptions, the emissions trading systems included in this report do not cover emissions from road transport.

Looking at the full distribution of ECRs confirms that emissions in the road sector are priced significantly more strongly than emissions in the non-road sectors (Figure 4.5). Figure 4.5 sorts CO_2 emissions from energy use according to the ECR at which they are priced, starting at zero. The horizontal axis shows the proportion of CO_2 emissions in the road or non-road sectors and the vertical axis shows the corresponding ECR (in EUR per tonne of CO_2).

The share of emissions priced at any given ECR in the road and non-road sectors can be read from Figure 4.5. For example, 54% or road transport emissions are priced at less than EUR 30 per tonne of CO_2, and 96% of non-road emissions. Also shown, 71% of road transport emissions face an effective carbon rate below EUR 90 per tonne of CO_2 while

100% of non-road emissions are priced below that level. The light blue price line, for road emissions, is very steep once the line at EUR 30 is crossed, indicating that a larger part of emissions is priced at much higher levels than EUR 30 per tonne of CO_2 than is the case for non-road emissions (dark blue line).

Figure 4.5. **Proportion of CO_2 emissions from energy use subject to different levels of effective carbon rates (biomass emissions included)**

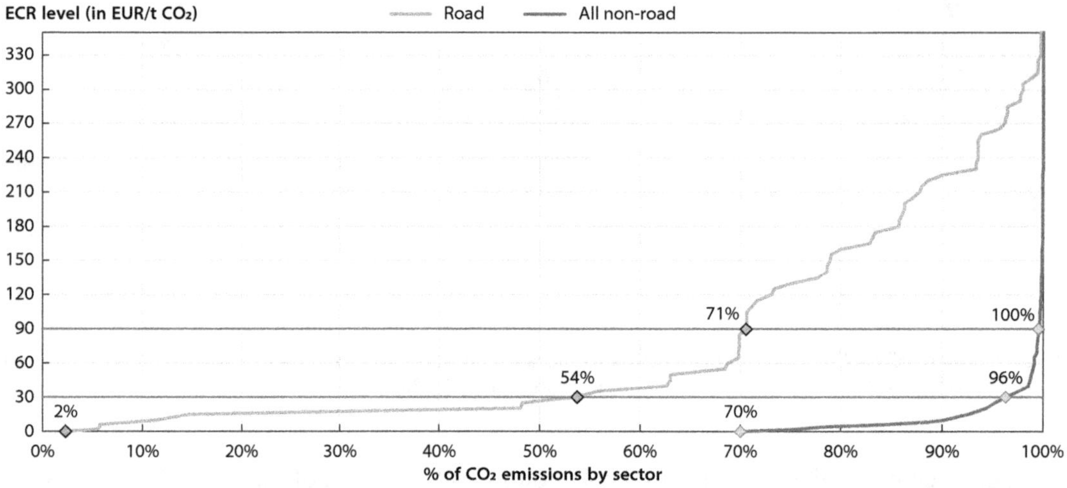

For road transport, specific taxes on energy use (namely excise taxes on gasoline and diesel) provide an incentive to reduce fuel consumption, and therefore CO_2 emissions, even though they may not have been introduced with the direct intention of abating such emissions. The higher rates on emissions in the road sector may be explained by the broader range of policy goals that governments often seek to address when taxing transport fuels. Firstly, energy use in road transport has a number of other negative external costs, such as local air pollution, congestion, accidents and noise. Countries may therefore tax transport fuels to make energy users consider these external costs in their transport choices, even though some of these are only very indirectly linked to fuel use and may be more effectively addressed by driving-based charges. Secondly, countries often tax transport fuels to raise revenue for the general budget or specifically to fund road infrastructure.

ECRs by country

ECRs vary strongly within and across countries, but there are similarities too. The higher ECRs observed in road transport, relative to non-road sectors, at the level of the 41 countries are also identified when looking at countries individually. Almost all CO_2 emissions from road transport energy are priced at or above EUR 30 in most countries, whereas emissions from energy use in non-road sectors rarely face an effective carbon rate above EUR 30 per tonne of CO_2. Further, a significant share of non-road emissions are not priced in many countries: twenty-one countries, for instance, do not apply any price to more than two-thirds of emissions from non-road sectors.

Figure 4.6 shows countries' CO_2 emissions from energy use by ECR interval, using the same ECR intervals as in Figure 4.4 and also distinguishing between emissions in road transport (left panel) and in all non-road sectors (right panel). In 2012, thirty-three countries priced at least 90% of road transport emissions at or above EUR 30 per tonne of CO_2 and only Brazil, Indonesia, Mexico (which has significantly increased road transport fuel taxes since), Russia and the United States priced all road transport emissions below EUR 30 per tonne of CO_2. For emissions in the non-road sectors, only the Netherlands priced more than 50% of emissions at or above EUR 30 per tonne of CO_2, and 40 countries priced less than 50% of non-road emissions at EUR 30 or more (of which seven priced no emissions from non-road at EUR 30, namely Brazil, Chile, India, Indonesia, Mexico, Russia and the United States). In the road sector, almost all emissions faced an ECR greater than zero, ranging from 81% of emissions in Brazil to nearly 100% coverage in 22 countries. In the non-road sector, the proportion of emissions covered by an ECR of greater than zero ranged from 0% in Chile and Indonesia to just over 90% in Italy, Israel and Korea.

A carbon price of at least EUR 30 per tonne of CO_2 covering all emissions in each sector would ensure that incentives for CO_2 emission abatement were consistent with the lower-end estimate of the climate cost of carbon. Lower prices may be a useful first step toward this goal, generating revenue and providing some incentive for abatement. To consider lower ECR levels, Figure 4.6 also shows the share of emissions priced at an ECR of EUR 5 per tonne of CO_2 or greater. Almost all CO_2 emissions in road transport are subject to an ECR

Figure 4.6. **Proportion of CO_2 emissions at different ECR intervals by country (biomass emissions included)**

The composition of effective carbon rates

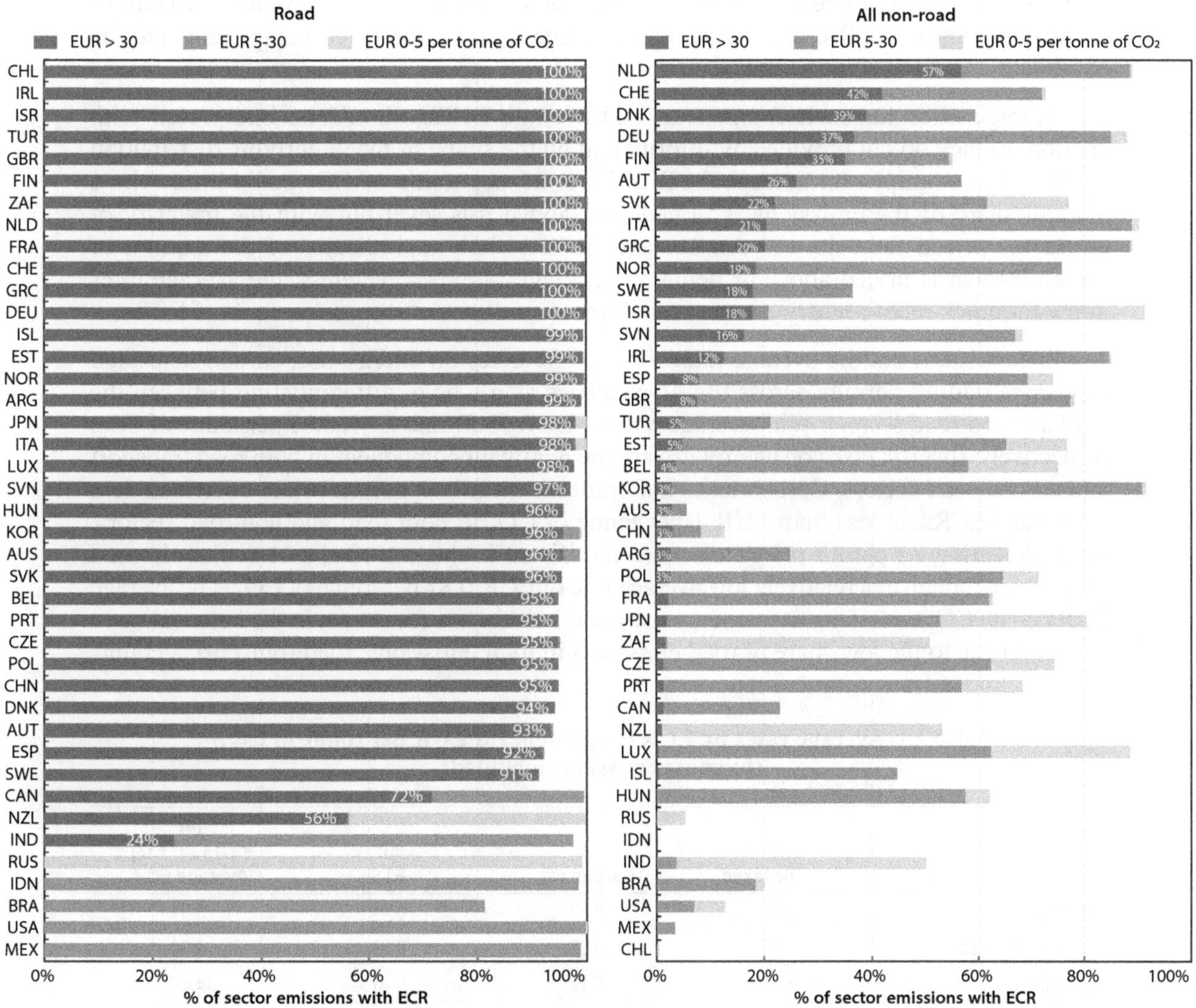

of above EUR 5 per tonne of CO_2, indicating that there is some policy-driven incentive for emissions reductions. In non-road sectors, the proportion of emissions priced at EUR 5 per tonne of CO_2 varies strongly between countries. Some countries do not price any non-road emissions at or above EUR 5 per tonne of CO_2 (Chile, Indonesia, and Russia), while twenty-four countries price at least 50% of non-road CO_2 emissions at or above this level.

The three market-based instruments used to price CO_2 emissions from energy use are employed in different ways across countries, applying to different (and sometimes overlapping) emission bases, and at widely varying rates. This section describes the contribution of each component (carbon taxes, specific taxes on energy use and tradable emission permit prices) to the levels and coverage of the ECRs observed in the 41 countries, distinguishing between road transport and the non-road sectors.

Average rates of ECR components

The average rate of each of the three ECR components is a summary measure that combines information on the rates and coverage of each price instrument. Average rates are weighted by the share of CO_2 emissions covered at each rate by each instrument. CO_2 emissions that do not face a price are therefore included in the denominator but not in the numerator used to calculate the averages.[1] Average rates show how much emitters pay on average to emit one tonne of CO_2 in a country or sector, and are a summary measure of the distribution of ECR components. The average rates of the three components can be combined into an average ECR.

Average rates for each component, and average ECRs, should be interpreted with caution, as they do not provide any information on the shape of the underlying distribution of rates. For example, an average ECR of EUR 25 per tonne of CO_2 could arise where a small share of the emissions is priced well above this level but with the majority of emissions priced far below it. An average ECR is thus no indicator of the amount of emissions that is priced above or below the average rate. Nevertheless, the average rates allow for quick inter-country comparison of the overall use of the components of ECRs.

Table 4.1 sets out the average rates for specific taxes on energy use, carbon taxes and emissions trading permits across the 41 countries as a whole, differentiating between the road and non-road sectors. With average rates of EUR 72.31 and EUR 3.05 per tonne of CO_2, respectively, specific taxes on energy use are the dominant component in both road transport and all non-road sectors. Carbon taxes and tradable emissions permit prices contribute less to average ECRs, at less than EUR 1 per tonne of CO_2 in both road and non-road sectors. Tradable emissions permit prices are more significant in the non-road sectors than in road transport, although at a relatively low average rate of EUR 0.87 per tonne of CO_2. The average ECRs on all CO_2 emissions from energy use are relatively low (as shown in the bottom line of Table 4.1) due to the low share of road emissions in total emissions: road transport accounts

Table 4.1. **Average rates by ECR component (in EUR per tonne of CO_2) (biomass emissions included)**

	(1) Specific taxes	(2) Carbon tax	(3) ETS	(4) = (1) + (2) + (3) Average ECR
Road transport	72.31	0.87	0.44	73.61
All non-road sectors	3.05	0.08	0.87	4.00
All sectors	13.44	0.19	0.81	14.44

for 15% of total CO_2 emissions from energy use in the 41 countries, giving road emissions a low weight in the average rates across all sectors.

A similar dominance of the specific tax component on energy use is seen within most of the 41 countries, although there is considerable variation in the level of average ECRs in each country and in the way the different price instruments are applied.

Figure 4.7 presents average ECRs in each country, disaggregated into the three price instruments, showing both emissions in road transport (left panel) and in non-road sectors (middle and right panel; note the difference in scale of the horizontal axis between the left- and rightmost panels). In road transport, specific taxes on energy use are by far the price instrument with the highest average rate in all countries except Sweden. Carbon taxes play a role in pricing road emissions only in the Nordic countries (Sweden, Norway, Finland, Denmark and Iceland) and in Ireland.

The high contribution of carbon taxes to average ECRs in road transport in Sweden is due to a relatively high general level of carbon pricing and also to the fact that the introduction of the Swedish carbon tax was accompanied by a commensurate reduction in

Figure 4.7. **Average effective carbon rates by country and price instrument (biomass emissions included)**

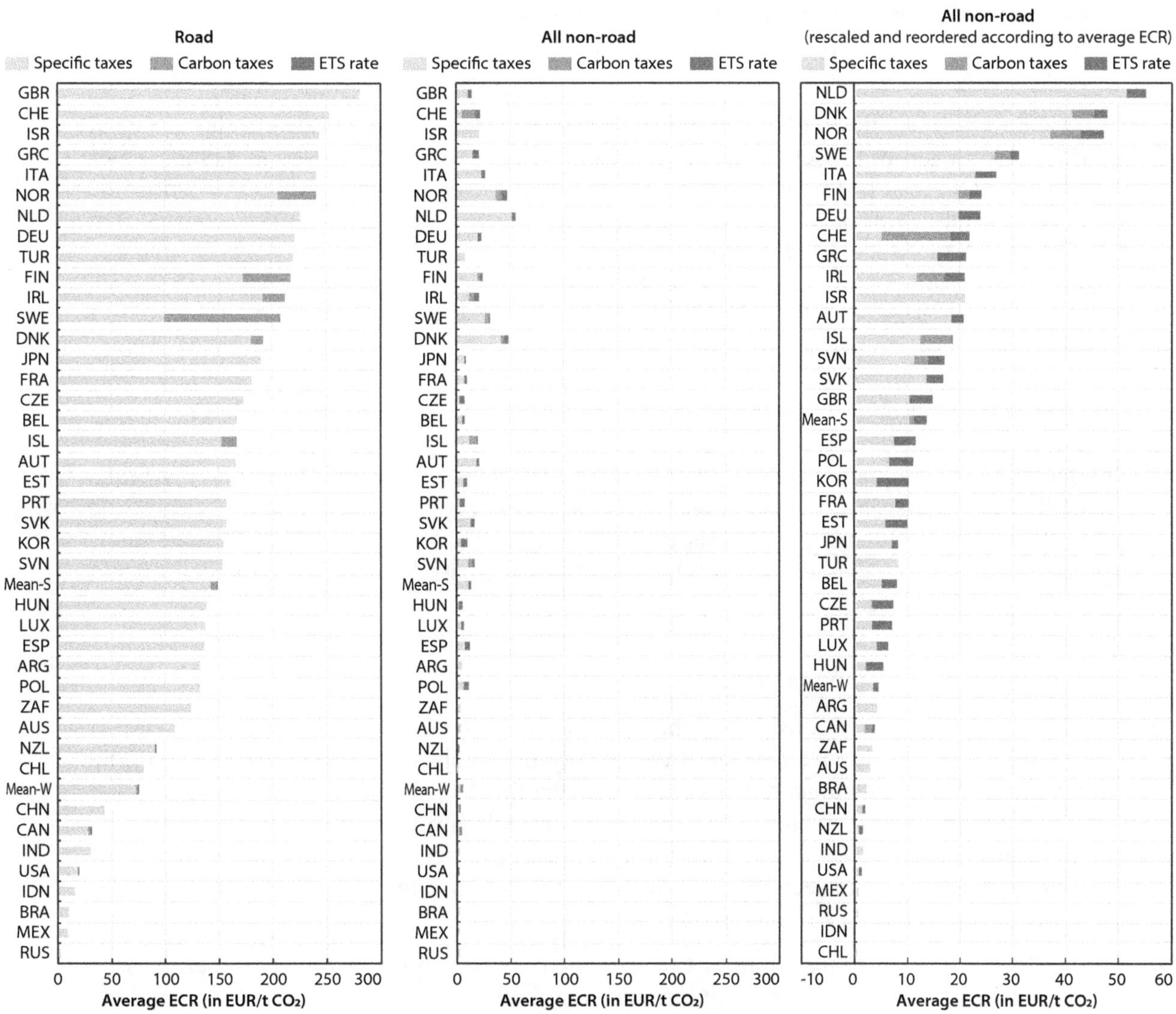

excise taxes. A similar approach was recently followed in France in 2014 (not reflected in the data shown in this report), where the excise tax rate for gasoline was reduced in parallel to the introduction of a carbon tax, although future increases in the carbon tax rate will not be accompanied by excise reductions. The cases of Sweden and France demonstrate that an understanding of carbon pricing requires consideration of all market-based instruments: considering one ECR component in isolation can give a misleading picture of how countries price CO_2 emissions if there are large differences in other components. Furthermore, considering policy change also requires looking at the three components, as an increase in one component can be accompanied by a reduction in another.

The average ECR on road transport emissions is above EUR 50 per tonne of CO_2 in all countries except China, Canada, India, the United States, Indonesia, Brazil, Mexico and Russia. The simple average ECR on road transport emissions across countries is above EUR 150, and is much higher than the weighted country average of less than EUR 100, indicating that some countries with high levels of CO_2 emissions have relatively low average ECRs in road transport.

The average ECR on emissions from non-road sectors (middle and right panels of Figure 4.7) is above EUR 30 per tonne of CO_2 in just four countries – Sweden, Norway and Denmark, and the Netherlands (the only country with a rate above EUR 50). Average ECRs on non-road emissions are above EUR 20 in seven other European countries (Austria, Ireland, Greece, Switzerland, Germany, Finland and Italy) and Israel. The other countries price non-road emissions at very low levels, in some cases applying a price close to zero.

The composition of ECRs is much more diverse in non-road sectors than in road transport. In non-road sectors, specific taxes on energy use are deployed to a much lesser extent than in road transport. Carbon taxes and tradable emission permit prices contribute more strongly to average ECRs in the non-road sectors, particularly in countries that historically apply specific taxes on energy use to a limited extent. For example, tradable emission permit prices make up more than 40% of the average ECR in non-road sectors in the Czech Republic, Estonia, Hungary, Korea, New Zealand, Poland, Portugal and the United States (even if the latter has no country-wide trading system). Carbon taxes represent more than 40% of the average non-road ECR in Switzerland, where they cover most residential and commercial emissions, and in Canada, where British Columbia's carbon tax applies at a relatively high rate (covering 9% of Canadian non-road emissions). Here too, the simple country average ECR is well above the weighted country average ECR. Low ECRs on non-road transport emissions suggest a potentially significant role for the carbon tax and emissions trading components of ECRs in pulling up average rates, in line with policy dynamics pointing towards an increased use of these explicit carbon pricing mechanisms.

Coverage of ECR components

A different way of considering the contribution of each price instrument to ECRs is to consider the share of emissions from energy use subject to each instrument, i.e. the coverage of each instrument. The coverage of the three instruments may overlap, so that the overall proportion of emissions covered by an ECR is often lower than the sum of the emissions covered by each of the three individual components. For example, an excise tax may apply to the same sector as a carbon tax; or an excise tax and an emissions trading system may apply to the same industry.

Table 4.2 presents the share of CO_2 emissions from energy use subject to each price instrument. It also shows the combined coverage of both tax instruments. The combined coverage of all three instruments is reported in the last column showing the proportion of emissions in each sector that face a positive ECR. As stated in the first section of this chapter, while 98% of road transport emissions are priced at a positive effective carbon rate, only 30% of emissions in other sectors are. Across all emissions from energy use only 40% are subject to a positive effective carbon rate.

Table 4.2. **Proportion of emissions from energy use subject to a positive effective carbon rate by price instrument (biomass emissions included)**

		Taxes			
	ETS	Carbon tax	Specific taxes	Combined tax	ECR
Road	5%	2%	98%	98%	98%
All non-road	13%	0%	23%	23%	30%
All	12%	1%	34%	34%	40%

Overall, carbon taxes are the smallest component of ECRs, followed by ETS prices. Specific taxes on energy use are much higher than both carbon taxes and ETS prices. The high ECR coverage in road transport is nearly entirely driven by the high coverage of specific taxes on energy use (98%), but the breakdown of coverage by price instruments in non-road sectors shows a different pattern, with specific taxes on energy use covering 23% of CO_2 emissions and tradable emissions permit prices covering 13%.

Price instruments can overlap, so that implementing new carbon pricing instruments does not necessarily increase the overall proportion of CO_2 emissions subject to a price. Where a tax applies to part of the same base as an ETS, this increases the price of emissions, but does not increase the total abatement under the ETS (in the presence of a binding cap), leading to lower permit prices under the ETS. This can reduce cost-effectiveness, as prices are no longer uniform. In Table 4.2, summing the coverage by individual price instruments can produce a number greater than the total sector coverage in the last column. For example, in road transport, across the 41 countries, nearly all emissions covered by an ETS are also subject to a specific tax on energy use. Similarly, carbon taxes and specific taxes on energy use often cover the same emissions in both the road and non-road sectors. The effect of carbon taxes on increasing coverage is therefore small. However, in non-road sectors, emissions trading systems more strongly contribute to total coverage. Taxes cover 23% of emissions, and this increases to 30% if trading systems are accounted for (an increase by 30%). Since emissions trading systems cover 13% emissions, there is a significant overlap between taxes and trading systems, as is discussed in more detail below and in Annex A.

Effective carbon rates within the non-road sectors

Emissions from non-road sectors account for 85% of total CO_2 emissions from energy use in the 41 OECD and selected partner economies. As seen above, they are less frequently subject to an ECR, and, on average, are subject to lower ECRs than road transport emissions. This section analyses the different non-road sectors in more detail. The non-road sectors include offroad transport, industry, agriculture and fishing, residential and commercial energy use, and the electricity sector. Of these, energy use in industry, electricity and in the residential and commercial sector gives rise to the largest proportion of carbon emissions (roughly 33% in the case of industry and electricity, and 15% in the case of the residential and commercial sector). Table 4.3 the share of CO_2 emissions from energy use from the different sectors.

Table 4.3. **Categorisation of CO_2 emissions from energy use**

Sector		CO_2 emissions included in sector	Proportion of economy-wide CO_2 emissions in all countries on average
	Road	All CO_2 emissions from energy used in road transport.	15.0%
Non-road sectors	**Offroad**	All CO_2 emissions from energy used in non-road transport (incl. pipelines, rail transport, domestic aviation and maritime transport).	2.2%
	Industry	All CO_2 emissions from energy used in industrial processes, in heating (incl. inside industrial installations) and in the transformation of energy, including fuels used for auto-generation of electricity in industrial installations.	33.3%
	Agriculture & fisheries	CO_2 emissions from energy used in agriculture, fisheries and forestry. CO_2 emissions from energy used in on-road transport are included in the road sector.	1.2%
	Residential & commercial	All CO_2 emissions from energy used for commercial and residential heating, including fuels used for auto-generation of electricity.	15.1%
	Electricity	All CO_2 emissions from energy used to generate electricity for domestic use excluding fuels used in the auto-generation of electricity.	33.2%

Non-road sector ECRs across OECD and selected partner economies

The proportions of emissions priced at different ECR intervals are broadly similar for the industrial, the electricity and the residential and commercial sectors when looking at the 41 OECD and selected partner economies as a group. As shown in the first row of Figure 4.8, the majority of emissions in each of these sectors (74% in the industry sector, 64% in the electricity sector and 80% in the residential and commercial sector) do not face a price and only a very small proportion is priced above EUR 30 per tonne of CO_2 (2%, 3% and 4% respectively). Between 5% and 13% of CO_2 emissions in these sectors are subject to an effective carbon rate between EUR 0 and EUR 5 per tonne of CO_2 and 11% to 19% to an effective carbon rate between EUR 5 and EUR 30 per tonne of CO_2.

Figure 4.8. **Proportion of CO_2 emissions subject to different ECR intervals by sector (biomass emissions included)**

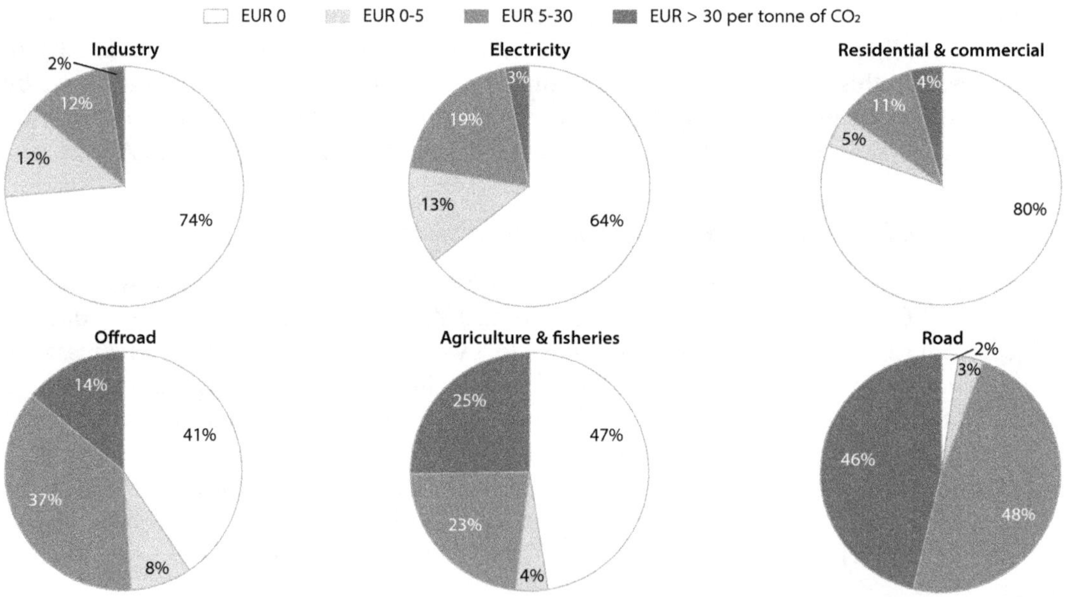

The shares of emissions within these same ECR intervals in offroad transport and in agriculture and fishing are similar (first and second panel in the second row of Figure 4.8). Compared to the other non-road sectors, a relatively high proportion of emissions from offroad transport and agriculture and fishing is priced, and prices are higher on average. Specifically, 41% and 47% of emissions in these sectors do not face a carbon price, 8% and 4% face a price between EUR 0 and EUR 5 per tonne of CO_2, 37% and 23% face a price between EUR 5 and EUR 30 per tonne of CO_2, and 14% and 25% face a price above EUR 30 per tonne of CO_2. For completeness, Figure 4.8 also shows the results for road transport, which have been described in previous sections.

Although exemptions and reduced rates are often applied to fuel use in non-road transport and in agriculture and fisheries (e.g. to fuel use in agriculture, or in rail transport), the rate structure at which these two sectors are taxed is more akin to the specific energy taxes applied on road use. The higher specific taxes on motor fuels for road use are often used as benchmark rates for the definition of the reduced rates (OECD 2015, 2013). In addition, energy use in these two sectors is dominated by oil products, which are often taxed at relatively high rates compared to coal or natural gas. Neither of the two sectors accounts for more than 2% of overall emissions across the 41 countries on average (Table 4.3), diminishing their weight in the analysis of ECRs. Given the very small proportion of emissions in these two sectors, the presentation and analysis of ECRs will from here focus on the ECRs applied to the remaining non-road sectors, which together comprise 81% of total emissions.

Non-road sector ECRs by country

Table 4.4 allows a closer examination of the pricing patterns of CO_2 emissions in the industrial, the electricity and the residential and commercial sectors (shown in the first row of Figure 4.8). These sectors account for the largest share of emissions on average in each country. The table shows for each country both the proportion of CO_2 emissions from energy use priced at a positive ECR or above EUR 30 per tonne of CO_2 in these sectors.

Patterns of ECR coverage and levels across the industrial sector and the residential and commercial sector are similar within the 41 economies: most countries price only a small proportion of emissions in these sectors at an ECR above EUR 30 per tonne of CO_2 and many countries do not at apply an ECR above EUR 30 to any emissions in these sectors. When looking at countries individually, some differences appear between both sectors. For example, Denmark, Italy, Israel, the Netherlands and Switzerland price a much higher proportion of emissions above EUR 30 per tonne of CO_2 in the residential and commercial sector than in the industry sector. Conversely, in Estonia, Hungary, Indonesia, Norway, Poland, Spain and the United Kingdom, a much larger share of emissions is subject to a positive effective carbon rate in industry than in the residential and commercial sector.

Countries that impose lower ECRs on industrial use, may be seeking to address competitiveness concerns, whereas countries that impose lower rates on the residential sector may place greater weight on concerns regarding the ability of low-income families to afford heating fuels. The lower rates may also be a result of different historical reasons for the imposition of taxes or trading systems on energy used in each sector. Regardless, where lower rates are intended to address competitiveness or distributional concerns, OECD work discussed in Chapter 2 has highlighted that the impact of pricing instruments on both of these factors is often overstated, and that it is generally preferable and possible to find ways to assist these sectors or households that do not weaken the price signal from energy pricing instruments.

Table 4.4. **Proportion of CO_2 emissions priced above EUR 0 and EUR 30 per tonne of CO_2 by country (biomass emissions included), percentages**

	Industry: Proportion of emissions priced at or above		**Electricity**: Proportion of emissions priced at or above		**Residential & commercial**: Proportion of emissions priced at or above	
	EUR 0	**EUR 30**	**EUR 0**	**EUR 30**	**EUR 0**	**EUR 30**
AUS	8	8	0	0	14	9
AUT	53	21	100	64	50	23
BEL	66	0	100	18	79	0
CAN	18	1	1	0	27	1
CHL	0	0	0	0	1	0
CZE	71	0	97	0	35	0
DNK	49	13	100	100	43	43
EST	73	4	100	0	16	6
FIN	49	25	100	88	30	30
FRA	78	1	100	15	38	0
DEU	80	11	100	82	80	1
GRC	92	29	100	21	61	9
HUN	78	0	100	0	23	0
ISL	19	0	100	100	26	0
IRL	76	15	96	0	74	21
ITA	88	5	100	0	80	61
ISR	35	5	99	13	87	49
JPN	62	1	93	2	100	0
KOR	92	2	92	0	95	20
LUX	80	0	100	0	95	0
MEX	4	0	1	0	1	0
NLD	77	32	100	100	95	59
NZL	44	1	54	0	57	3
NOR	84	12	100	100	23	21
POL	76	2	100	0	9	0
PRT	59	0	100	0	27	6
SVK	70	20	100	0	90	31
SVN	79	9	100	0	36	27
ESP	76	3	100	0	25	20
SWE	33	10	100	100	21	21
CHE	64	17	7	7	82	58
TUR	43	3	100	0	29	0
GBR	84	14	100	0	33	10
USA	5	0	8	0	8	0
ARG	59	1	68	0	94	5
BRA	13	0	100	0	0	0
CHN	13	3	13	0	5	4
IND	55	0	81	0	6	0
IDN	0	0	0	0	0	0
RUS	5	0	0	0	8	0
ZAF	8	2	100	0	2	1

In the electricity sector, the majority of countries apply an ECR of greater than zero to a large proportion of CO_2 emissions or even the entire emissions base. Twenty-nine countries price more than 90% of carbon emissions from the industry sector and twenty-two countries price all emissions from the electricity sector (through combinations of emissions trading systems and taxes on electricity use). Only Australia, Chile, Indonesia and Russia, do not price any emissions from electricity generation. The proportion of emissions in the electricity sector subject to an ECR of greater than EUR 30 per tonne of CO_2 differs significantly across countries: eight countries price more than 50% of emissions from the generation of electricity above EUR 30 per tonne of CO_2, while twenty-four countries do not price any of the emissions from the fuels used to generate electricity above EUR 30 per tonne of CO_2.[2]

Impact of different price instruments in the non-road sectors

The impact of different carbon pricing instruments on the level and coverage of ECRs differs among the three non-road sectors analysed. Figure 4.9 presents the average rate for specific taxes on energy use, carbon taxes, and tradable emission permit prices in industry, the residential and commercial sector, and in electricity generation.

Figure 4.9. **Specific taxes dominate ECRs in non-road sectors (biomass emissions included)**

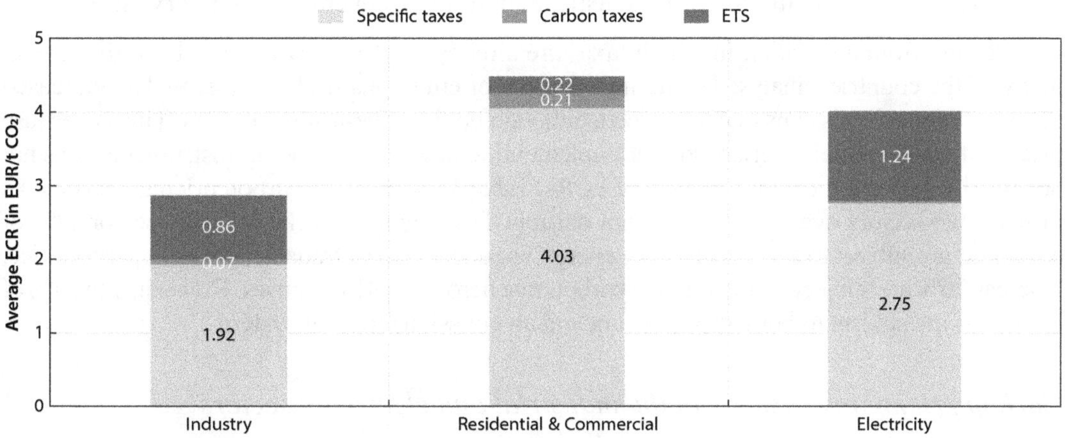

Specific taxes on energy use are the largest component of the average ECR in each of the three non-road sectors shown here. The relative contribution of tradable emissions permits to overall ECRs is relatively large in industry and in electricity. Carbon tax rates are relatively low on average in all sectors under consideration. Average ECRs in the industry and the electricity sector are lower than in the residential and commercial sector, also because of the smaller contribution of excise taxes applying to fuels used by these sectors. The composition of the average ECR in the residential and commercial sector differs from the other two sectors, due to the larger influence of the specific taxes on energy use. The ECR component made up by tradable emissions permits is smaller in the residential and commercial sector, since ETS tend to cover mostly larger firms.

Average ECRs provide a first comparison of the relative importance of the different price instruments, combining both rate and coverage information into a single metric. However, they do not directly show the proportion of emissions covered by each instrument. Table 4.5 supplements the information shown in Figure 4.9 by reporting the share of CO_2 emissions covered by each instrument at a positive rate, as well as the combined total tax coverage. The combined coverage of all three instruments in the last column shows the proportion of emissions in each sector that is subject to an ECR.

Table 4.5. **Proportion of emissions subject to a positive effective carbon rate by price instrument (biomass emissions included)**

		Taxes			
	ETS	Carbon tax	Specific taxes	Combined tax	ECR
Industry	13%	0%	17%	17%	26%
Residential & commercial	3%	1%	17%	18%	20%
Electricity	18%	0%	27%	27%	36%

Of the three large non-road sectors analysed here, ECRs cover a maximum of 36% of emissions in the electricity sector, and lower shares in the industrial (26%) and the residential and commercial sector (20%). In the industrial and electricity sectors, emissions trading systems are much more influential in terms of their coverage, at 13% and 18% respectively, whereas specific taxes on energy use are less important in terms of coverage at 17% of CO_2 emissions in industry and 27% in electricity. The overlap between emissions trading systems and taxes is larger in electricity than in industry (it is smaller in residential and commercial use, because of the small coverage of trading systems). Carbon tax coverage is low in both sectors, and overlaps with the coverage of specific energy taxes, implying that the contribution of this instrument to ECR coverage is not very large.

Starting from a situation in which taxes are already in place – which has been the case in many of the countries analysed – the introduction of emissions trading systems has increased the proportion of emissions that are priced, although not in a one-to-one relation. This is because price instruments often overlap, so that implementing new carbon pricing instruments does not necessarily increase the total coverage of ECRs. Table 4.5 shows that carbon pricing instruments in non-road sectors overlap, although not entirely. For example, in the industry sector 17% of emissions are subject to a specific tax on energy use and 13% to a tradable emission permit price, whereas 26% are subject to an effective carbon rate across the 41 countries. Consequently, 4% of emissions are subject to both a specific tax and an emissions trading system.

Carbon pricing instruments in the industrial and electricity sectors

The importance of carbon pricing instruments in determining ECRs varies between sectors, as seen in the previous subsection. While specific taxes on energy use are the most influential in constituting ECRs in all sectors, emissions trading systems are of particular relevance to the pricing of carbon emissions in the industrial and electricity sectors. Almost all of the emissions trading systems apply most strongly, in terms of the emissions subject to the system, to these two sectors. Of the six sectors analysed here, the industry and electricity sector are also the largest in the share of CO_2 emissions from energy use, each at just over 33% each for the 41 economies as a whole.

In the industry sector, emissions trading systems apply to 13% of CO_2 emissions from energy use across the 41 countries, as shown in Table 4.5. This relatively high coverage (compared to the extent to which emissions trading systems apply in most other sectors) causes the comparatively greater influence of emissions trading systems in determining average ECRs in the industrial sector, as shown in Figure 4.9 (though the contribution of specific taxes on energy use to ECRs remains higher).

Within the 41 countries, the influence of emissions trading systems in determining ECRs varies. Twelve of the countries considered do not have an emissions trading system. In the countries which price emissions via tradable permits, the coverage of emissions trading systems in the industrial sector varies from over 90% in Greece and Korea to 1% in Luxembourg and 5% in China. By way of contrast, taxes (including both specific taxes on energy use and carbon taxes) apply to industrial emissions in all countries except Chile, Indonesia and the United States, covering on average 37% of emissions in the remaining countries. The degree of overlap between the two instruments (i.e. the share of emissions that are subject to both emissions trading systems and taxes) ranges from 9% in Portugal to 57% in Greece.

To highlight the differences in the application of the different price instruments across countries, Figures 4.10 and 4.11 compare the contribution of taxes and emissions trading systems to ECR coverage and average ECRs in the industrial sector in each country. Figure 4.10 shows the proportion of emissions subject to a positive ECR, distinguishing between CO_2 emissions that are covered only by taxes (including carbon taxes and specific taxes on energy use), emissions that are covered only by a tradable emission permit price, and emissions that are covered by both by taxes and emissions trading systems.[3] Figure 4.11 presents average ECRs in the industry sector, disaggregated by price instrument, excluding unpriced emissions. The order of countries is the same in both figures to allow comparison between the proportion of emissions subject to each price instrument and each instrument's average rate in each country.

Figure 4.10. **Proportion of emissions subject to a positive effective carbon rate in the industry sector (biomass emissions included)**

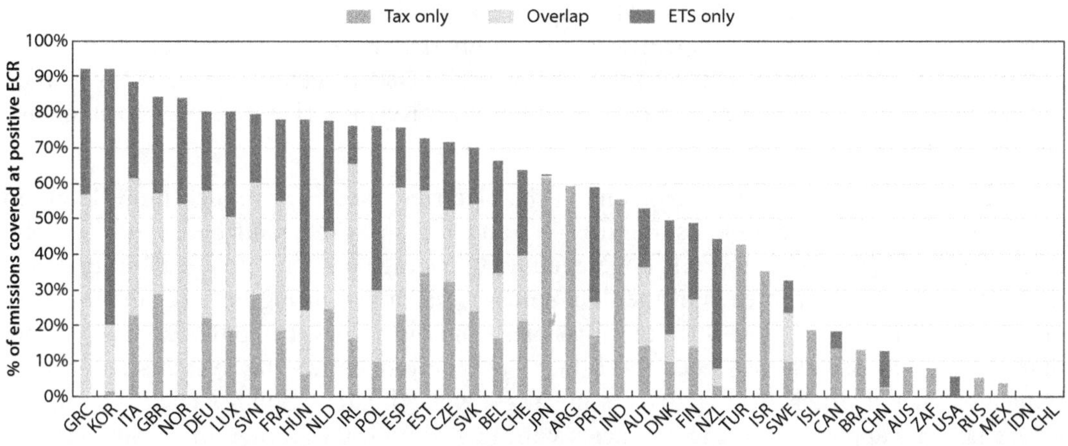

Figure 4.11. **Average effective carbon rates in the industry sector (biomass emissions included)**

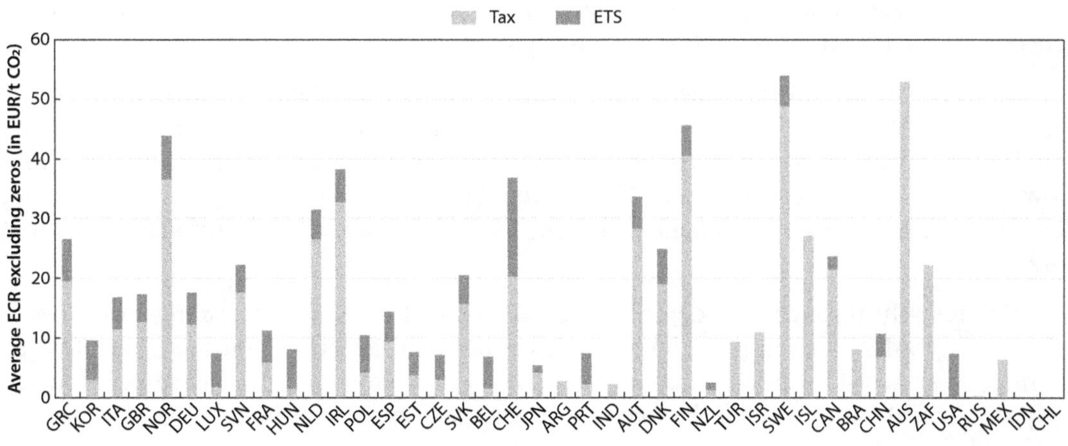

Average tax rates on emissions (here excluding zero rates) in the industrial sector are very heterogeneous across countries, ranging from EUR 54 per tonne of CO_2 in Sweden to very low rates in India, New Zealand and Russia and no ECR at all in Chile, Indonesia and the United States.[4] In most countries, taxes on CO_2 emissions from energy use remain the dominant instrument in their contribution to overall ECRs, although in of Belgium, the Czech Republic, Estonia, Hungary, Korea, Luxemburg, Poland, Portugal, and the United States, trading systems dominate. The United States is the only country with the entire average ECR constituted by price signals from emissions trading systems (California and RGGI), whereas in eight countries the ECR consists entirely of taxes on energy use.

Comparison of the proportion of emissions that are subject to a pricing instrument, and the average rates of each instrument, shows that emissions trading systems typically apply at lower rates than taxes on CO_2 emissions from energy use (Figure 4.11). However, in all countries where emissions trading systems apply, they increase the share of emissions covered at a positive rate in the industrial sector beyond those covered by taxes (Figure 4.10).

In the electricity sector, emissions trading systems apply to 18% of CO_2 emissions from energy use, as shown in Table 4.5. This is the sector in which emissions trading systems have the greatest coverage, which helps explain the comparatively larger average rate resulting from emissions trading systems in this sector (see Figure 4.9). However, the coverage of emissions trading systems in the electricity sector strongly overlaps with that of specific taxes on energy use, both at an aggregate level and within 22 of the 30 countries in which an ETS is applied. Specific taxes on energy used to generate electricity fail to distinguish between the carbon-intensity of fuels used to generate electricity if they are levied on an energy content basis, but they do raise the price of emission-intensive electricity all the same.

Within each country, CO_2 emissions in the electricity sector are much more likely to be priced than industry emissions. Figures 4.12 and 4.13 show the contribution of taxes and emissions trading systems to ECR coverage and to average carbon rates (excluding zeros) in the electricity sector in each of the 41 countries. As shown in Figure 4.12, the majority of countries price more than 90% of emissions from electricity. In countries where more than 90% of emissions are priced, both taxes and emissions trading systems are widely implemented and largely overlap (except in Brazil, Iceland, Israel, South Africa and Turkey, where no ETS applies). In some countries with low or no tax coverage in certain sectors, emissions trading systems contribute significantly to total coverage. For example, emissions trading systems increase the proportion of emissions subject to an ECR in the electricity sector from 27% to 92% in Korea and from 0% to 54% in New Zealand. In China and the United States, implementation of regional emissions trading systems broadens ECR coverage from 0% to 13% and 8% respectively.

Average ECRs on emissions in the electricity sector are very heterogeneous across countries, as shown in Figure 4.13. As in the industrial sector, emissions trading systems contribute significantly to average carbon rates in countries where tax rates are relatively low or zero, including China, the Czech Republic, Estonia, Hungary, Ireland, Korea, Luxemburg, New Zealand, Poland, Portugal, the Slovak Republic, Spain and the United States.

Figures 4.10 through 4.13 demonstrate that the different carbon pricing instruments do not only affect carbon price levels but also have strong effects on the proportion of emissions subject to a price. A comprehensive evaluation of pricing instruments hence requires consideration of both the rates and coverage of each instrument.

Figure 4.12. **Proportion of emissions subject to a positive effective carbon rate in the electricity sector (biomass emissions included)**

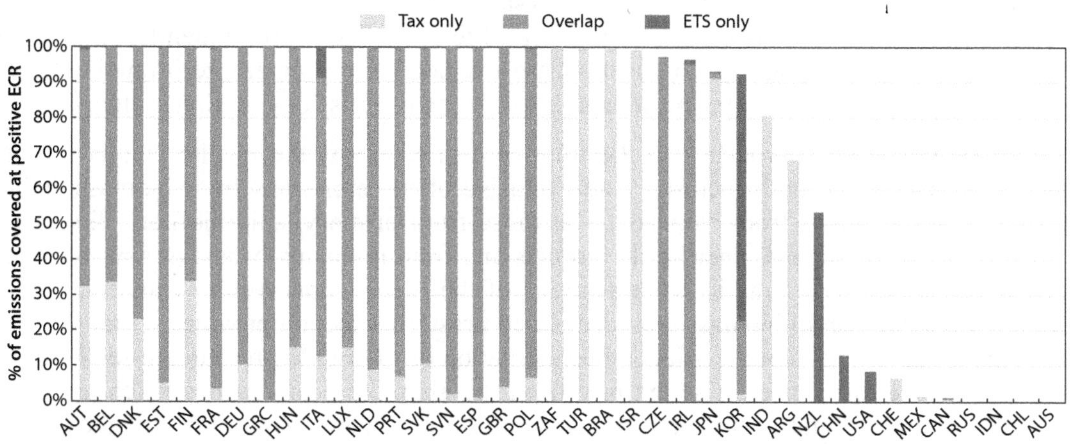

Note: The graph excludes three outliers – Iceland, Norway and Sweden.

Figure 4.13. **Average effective carbon rates in the electricity sector (biomass emissions included)**

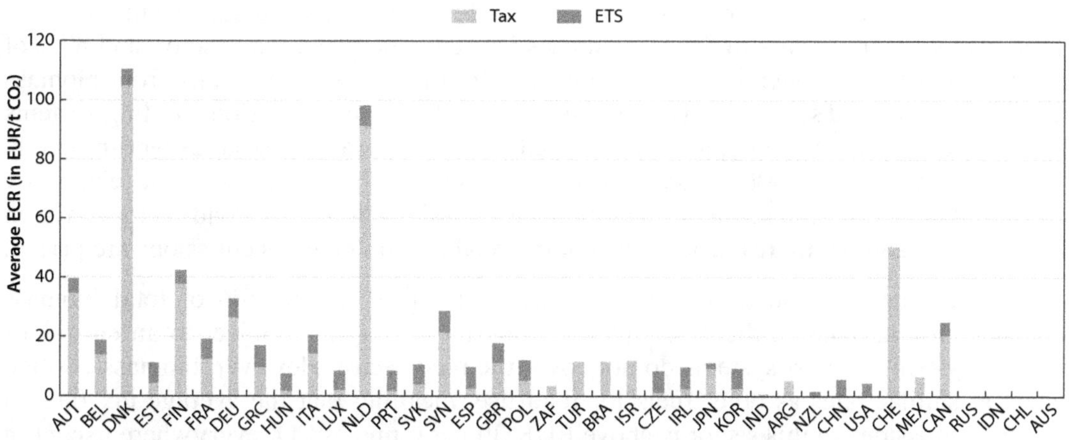

Note: The graph excludes three outliers – Iceland, Norway and Sweden.

Low carbon rates covering a large proportion of emissions may represent a first step towards pricing emissions and to making polluters pay for the climate cost of their emissions. Continued extension of carbon rate coverage to a wider set of emissions in the industry and electricity sectors – responsible together for 66.5% of all CO_2 emissions from energy use in the 41 countries – would strengthen carbon pricing policies. The choice of ECR component will largely depend on the individual circumstances and preferences of policy makers in each country.

Treatment of biomass in the calculations

The analysis of ECRs in the previous sections of this chapter does not differentiate between CO_2 emissions from the combustion of biomass and CO_2 emissions from the combustion of fossil fuels, consistent with the approach taken in previous OECD work

(OECD, 2013 and 2015). Total CO_2 emissions from energy use in each country in the results above hence include emissions from biomass combustion. However, many countries and the design of many emissions trading systems take a different approach to biomass emissions, treating them as carbon neutral on a life cycle basis. Consequently, emissions from biomass at the point of combustion are excluded from (or are zero-rated in) the consideration of CO_2 emissions from energy use in many emissions trading systems.

To capture both approaches to the treatment of biomass, the results in this report have been calculated both inclusive and exclusive of the emissions from biomass. This report takes emissions from biomass to be those from primary solid biofuels, biogases, biogasoline, biodiesel and other liquid biofuels and from municipal waste (renewable and non-renewable). The results in other sections of the report show results with emissions from the combustion of biomass included. In contrast, this section provides a snapshot of how the treatment of biomass affects the calculation of the ECRs presented elsewhere in the report. The differences arise entirely from the alternative approaches taken towards the accounting of emissions from biomass, and are not due to any policy changes. The degree to which there are differences in the results when biomass emissions are excluded from the analysis depends on three elements: firstly, the degree to which biomass is used in the country or sector being considered; secondly, the way in which biomass emissions are priced (either through taxes or through emissions trading systems); and thirdly, the distribution of ECRs on non-biomass emissions in the country or sector considered.

The use and pricing of biomass emissions differs between sectors, as summarised in Table 4.6: 95% of biomass in the 41 countries is used outside the road sector, and 89% of biomass is used in residential heating and industrial processes. Emissions from biomass used in the non-road sectors are typically not subject to a price. Emissions trading systems zero-rate or exclude biomass, and in non-road sectors it is rarely taxed, except indirectly when used to generate electricity (a very small proportion of biomass) via electricity consumption taxes. As a result, 94% of biomass emissions in the non-road sectors have an ECR of zero, and within residential and commercial use, no biomass emissions are priced.

Biomass use in road transport is limited and causes around 5% of total biomass combustion emissions in the 41 countries. Emissions rarely are subject to an emissions trading system, as most systems do not cover the road sector. However, biomass is often taxed at similar rates as other fuels for road use, meaning that the average tax rate on biomass emissions in this sector is above EUR 30 per tonne of CO_2 everywhere except in the United States. Also, the proportion unpriced biomass emissions is smaller in the road sector, at 36% across the 41 countries as a whole (although across countries, 100% being unpriced in 21 countries and all biomass emissions being priced in 11 countries).

Consequently, the impact of excluding biomass emissions from the calculations has different impacts in the road and non-road sectors. In the non-road sectors, excluding biomass emissions removes a sizeable proportion of emissions that are priced at zero, reducing the proportion of emissions with an ECR of 0 from 70% to 66% of total emissions. As few biomass emissions in non-road sectors are priced at a positive ECR, and almost none above EUR 30 per tonne of CO_2, the impacts on ECR coverage at higher rates are primarily due to the distribution of ECRs on non-biomass emissions in the non-road sectors. The overall average ECR in the road sector therefore increases only slightly when biomass is excluded, from EUR 4 to EUR 4.40 per tonne of CO_2.

In the road sector, only 36% of biomass is taxed at zero, and the use of biomass is smaller, so the impact of biomass treatment is smaller. Coverage changes from 2% of total emissions with biomass included to 1% with biomass excluded. Most biomass in the road sector is taxed above EUR 30 (both across and within countries), so that a 2 percentage-point increase in coverage is observed at ECRs between EUR 5 and EUR 30, and this is because the average ECR on biomass emissions in the United States is just under EUR 25 per tonne of CO_2. Excluding biomass emissions from the calculations increases the average ECR in the road sector from EUR 73.61 to EUR 74.45 per tonne of CO_2.

Table 4.6. **Characteristics of CO_2 emissions from biomass**

			Road	All non-road	All emissions
Biomass characteristics	Biomass as % of all emissions		4	13	12
	Unpriced biomass as % of all biomass emissions		36	94	91
	% of biomass used in sector		5	95	100
Coverage (% of total emissions)	ECR 0 (i.e. unpriced)	Biomass included	2	70	60
		Biomass excluded	1	66	56
	ECR 5	Biomass included	94	19	30
		Biomass excluded	96	21	33
	ECR 30	Biomass included	46	4	10
		Biomass excluded	47	4	11
Average ECR (EUR per tonne of CO_2)	Biomass included		73.61	4.00	14.44
	Biomass excluded		74.45	4.40	15.88

Within individual countries, most biomass use occurs in the non-road sectors at low or zero prices. The impact of excluding biomass emissions on ECR levels and coverage for different countries largely depends on the proportion of biomass emissions to total emissions in each country. Countries with a higher proportion of biomass use, particularly outside the road sectors, will see greater responsiveness of ECR results to the treatment of biomass emissions, particularly where other emissions are subject to higher prices. In ten of the 41 countries, more than 20% of total CO_2 emissions are from biomass combustion: Sweden, Finland, Brazil, Indonesia, Chile, Denmark, India, Austria, Portugal, and Estonia. In these ten countries, and in addition in Slovenia, Norway, France, Switzerland and New Zealand, more than 20% of emissions in the non-road sectors is from biomass combustion. In these countries, omitting emissions from biomass combustion leads to significantly higher ECRs. In Sweden, for example, the share of emissions priced at EUR 30 per tonne or more, increases from 34% to 60% if emissions from the combustion of biomass are not included.

Effective carbon rates: The bigger picture

The previous sections investigated ECR levels and coverage in the 41 countries and six sectors, on an aggregate level as well as country-by-country. This section analyses ECRs in a broader context. The "carbon pricing gap" is introduced in the first subsection ("The carbon pricing gap"). The second subsection ("A counterfactual scenario: increase prices and coverage to at least current median levels") uses the carbon pricing gap to simulate counterfactual scenarios, illustrating the impact of different pricing policies on ECR level and coverage. The third subsection ("Effective carbon rates and macroeconomic characteristics") relates the results of the analysis of ECRs to selected broader economic characteristics, i.e. total CO_2 emissions, gross domestic product (GDP), population size and net energy imports.

The "carbon pricing gap"

The extent to which the 41 countries price CO_2 emissions from energy use using market-based instruments can be measured by comparing current prices to the conservative cost estimate of EUR 30 per tonne of CO_2. Specifically, the extent to which emissions are priced at less than EUR 30 per tonne of CO_2 can be seen as the "carbon pricing gap" from market-based policy instruments. This gap here is measured as a percentage. If the ECR on all emissions was at least EUR 30 per tonne, the gap would be zero, and if the ECR was zero throughout, the gap would be 100%.

The indicator is illustrated in Figure 4.14 for the existing ECRs. The dashed line shows the current distribution of ECRs across the 41 economies as a whole. As indicated earlier, 60% of all emissions are not subject to an ECR, and only 10% are subject to an ECR of greater than EUR 30 per tonne of CO_2. The proportion of emissions below the dashed line is subject to a positive ECR. The proportion of emissions between the dashed line and the horizontal line at EUR 30 per tonne of CO_2 (shaded in dark blue in Figure 4.14) is either not priced by any price instrument, or is priced at a level below EUR 30 per tonne of CO_2. The area in dark blue is the carbon pricing gap across the 41 countries, and it equals 80.1%.

Figure 4.14. **The carbon pricing gap, shown in dark blue**

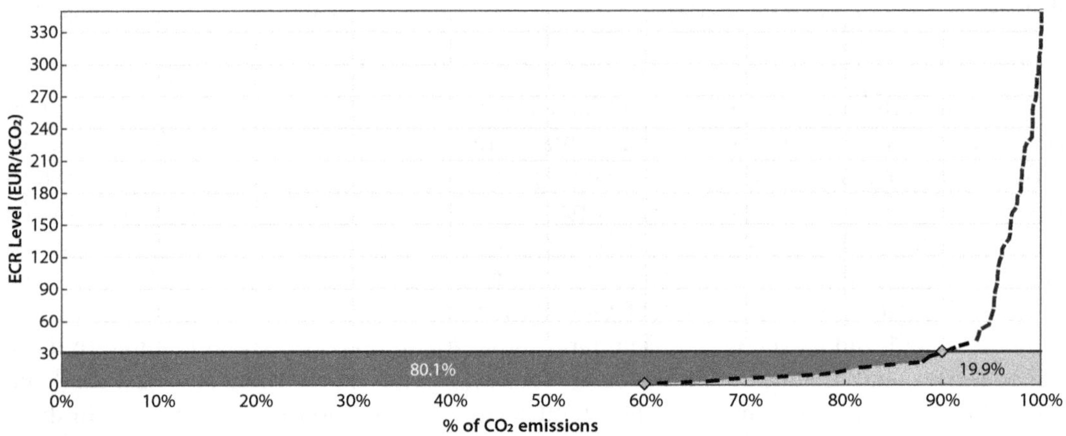

Figure 4.14 can be reshaped to focus solely on the carbon pricing gap, by cutting the vertical axis of the graph at the reference point (EUR 30 per tonne of CO_2), as in Figure 4.15 (which otherwise presents the same information as Figure 4.14). Again, focussing on EUR 30 per tonne of CO_2 does not imply that a carbon price at this level would be sufficient to internalise the climate cost of carbon and nor does it serve as a normative goal.

Figures 4.14 and 4.15 present the results for the group of 41 countries as a whole. From a climate perspective, it is the overall effect of carbon pricing which is important, rather than its location. However, within this group of countries, the individual country carbon pricing gaps are likely to vary substantially, reflecting the very different applications of carbon pricing among countries. In particular, some larger economies currently feature low ECRs, and this significantly increases the carbon pricing gap, due to their weight in overall emissions.

Figure 4.15. **Carbon pricing gap, focusing on priced and unpriced emissions below EUR 30**

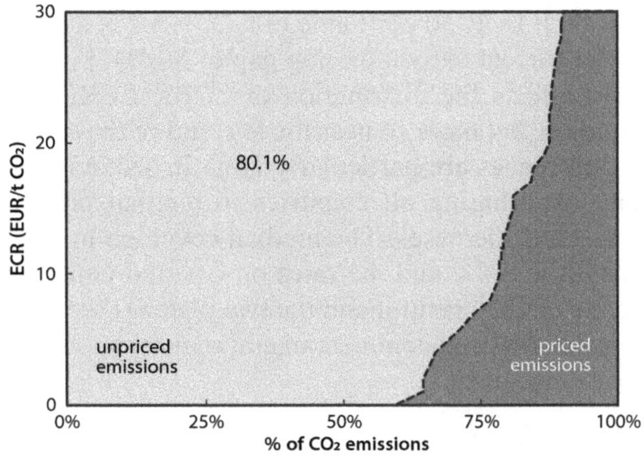

A counterfactual scenario: increase prices and coverage to at least current median levels

The carbon pricing gap presently equals 80.1% as shown above. If carbon pricing is to play a more significant role in climate policy, this gap needs to decline. Reductions can be achieved in at least two ways, first, by extending the share of emissions subject to a price and, second, by increasing the price level of either taxes or trading systems. Both avenues are under consideration in many of the countries considered, and are likely to be increasingly relevant following the Paris Agreement. For example, China is planning to implement a national ETS in 2017 (National Development and Reform Commission, 2016), which will significantly increase the global share of CO_2 emissions from energy use subject to a carbon price. In the European Union ETS, the low prices of tradable emissions permits triggered the development of a market stability reserve. This reserve will start operating as of 2018 (European Commission, 2016) and will remove surplus emissions permits from the market, which is expected to lead to higher carbon price levels. Increases in carbon pricing via taxes are also underway in a few countries. For example, a carbon tax is set to be levied in Chile and in the Canadian province of Alberta starting in 2017, and a draft carbon tax bill has been the subject of consultation with stakeholders in South Africa.

The analysis of ECRs has revealed that they are very often low and that there are large differences between countries, both on an economy-wide basis and also within the six sectors analysed. It follows that improving the performance of countries with low rates or low coverage in any of the sectors would reduce the carbon pricing gap. To make this notion more tangible, a counterfactual scenario was calculated. This scenario is constructed in two steps. First, the median coverage and the median ECR (for emissions covered by an ECR) are identified for each of the six sectors. (The median countries are different in all cases but one and are regionally and economically very diverse.) Second, for each country, the coverage and the ECR are increased to at least the median level. If coverage or ECR level already were higher than the median, they are left as is. In other words, the distribution of coverage and rates is modified by increasing countries at the low end to the median of the currently observed distribution in both coverage and rates. Energy use is kept constant in this exercise, so behavioural effects of the price changes are not accounted for. Since the counterfactual scenario involves price increases, energy use and carbon emissions in fact would decline as a response to higher prices.

The carbon pricing gap in the counterfactual scenario is 53.1%. This says that, if ECR rates and coverage in each of the six sectors would increase to the median levels for those emissions where observed rates and coverage are below the median, 53.1% of emissions would be priced at less than EUR 30 per tonne of CO_2. This is a decline of 27%-point, or of 34%, compared to the current carbon pricing gap of 80.1%. Figure 4.16 shows how the counterfactual scenario affects the distribution of ECRs. ECRs are higher everywhere than in the current situation, but more so near the low end of the distribution. This reflects the fact that country differences are particularly large in sectors with low ECRs, so that reducing heterogeneity by bringing all countries to median price and coverage levels leads to relatively large ECR increases. The median coverage in electricity is 100%, and for industrial emissions it is 59%, and the rates on covered emissions in both cases are around EUR 11 per tonne of CO_2, resulting in the two plateaus seen in Figure 4.16. Median coverage is lower for residential and commercial emissions (at 30%) but the minimum rate is just over EUR 20 per tonne of CO_2.

Figure 4.16. **Carbon pricing gap under counterfactual scenario of median prices and coverage**

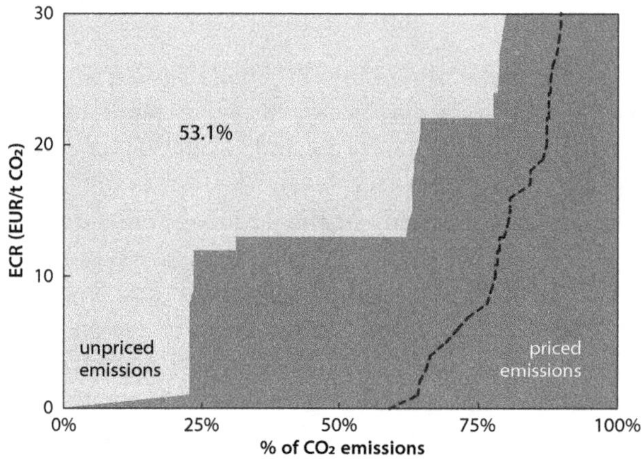

To sum up, the counterfactual scenario illustrates that "catching up" with median performance results in strong reductions of the carbon pricing gap. Median performance is not obviously related to country characteristics including GDP, energy use and carbon intensity of energy. While median performance may not be equally easy to attain for every country, it is not necessarily out of reach because of its particularly high level of ambition either.

Effective carbon rates and macroeconomic characteristics

ECR levels and the share of emissions that are priced at different rates vary substantially across countries, as the previous sections have shown. To shed some light on factors that potentially help explain these differences, this section relates, for each country, the level and coverage of ECRs to four macroeconomic characteristics of each country: total CO_2 emissions, gross domestic product (GDP), population size and net energy imports.

Figure 4.17 plots the carbon intensity of the economy in each country against ECR coverage in that country. The left panel shows the proportion of emissions facing an ECR at or above EUR 30 per tonne of CO_2, and the right-hand panel depicts the proportion of emissions facing an ECR above EUR 0 per tonne of CO_2.

Figure 4.17 demonstrates that countries that price a high share of CO_2 emissions above EUR 30 per tonne of CO_2 tend to have a low carbon intensity of GDP (left panel). The same holds to a lesser degree for ECRs greater than zero (right panel). The right-hand panel also shows that a cluster of mostly European economies, together with Israel, Japan, and Turkey, display both a relatively high share of priced emissions and have relatively low carbon intensities of GDP, but there is more heterogeneity among countries with relatively high carbon intensities of GDP.

Figure 4.17. **Proportion of CO_2 emissions priced above EUR 30 (left) and EUR 0 (right) per tonne of CO_2 relative to the carbon intensity of GDP, 41 countries, 2012**

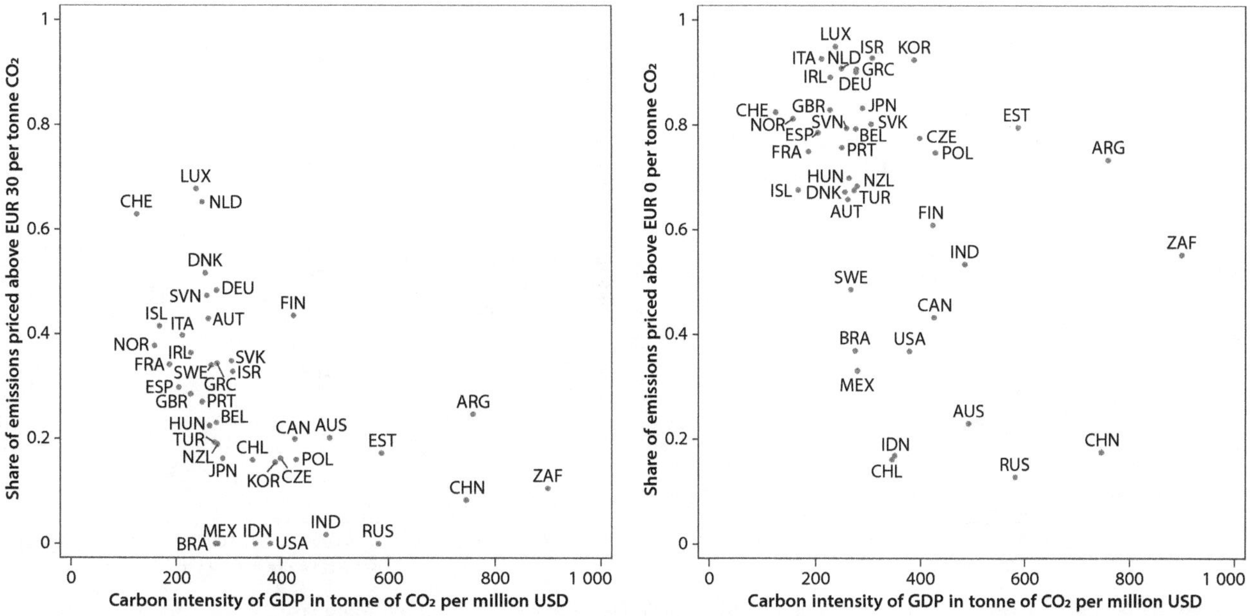

Source: GDP data is from the World Bank (2016), *World Development Indicators* (database), http://data.worldbank.org/data-catalog/world-development-indicators.

The inverse relationship between the proportion of emissions priced at different carbon rates and the carbon intensity of GDP does not necessarily imply a direct causal effect in either direction. A low carbon intensity of GDP may be the result of carbon pricing policies that reduce energy consumption or steer towards less emission-intensive energy sources. However, it may also be the case that countries with low carbon intensities of GDP find it easier to implement stricter carbon pricing policies. That said, there is strong empirical evidence that higher carbon prices discourage emissions, and reducing emission intensities is one way of doing so.

Figure 4.18 displays the same levels of ECR coverage (at EUR 30 and EUR 0, left and right panels respectively) against GDP per capita. It shows that increasing per capita income is associated with higher ECR coverage. Countries with higher per capita incomes tend to price greater shares of emissions above EUR 30 per tonne of CO_2 (left panel). Exemptions to this pattern occur in the United States, which prices a low share of emissions above EUR 30 per tonne of CO_2, despite its relatively high income levels, and Argentina, which prices a relatively high share of emissions above EUR 30 per tonne of CO_2 despite its relatively low income level.

The relationship between the share of emissions with an ECR above EUR 0 per tonne of CO_2 and GDP per capita is less pronounced (right-hand panel of Figure 4.18). Many countries located towards the middle of the income distribution price a significant share of emissions at least at some positive price. At the same time, some high income countries price a comparatively low share of emissions. Both panels of the figures show that, while high income countries are more likely to price a high share of emissions above EUR 30 per tonne of CO_2, many countries with a lower GDP per capita also implement carbon prices, though typically at lower levels.

Figure 4.18. **Proportion of CO_2 emissions priced above EUR 30 (left) and EUR 0 (right) per tonne of CO_2 relative to GDP per capita**

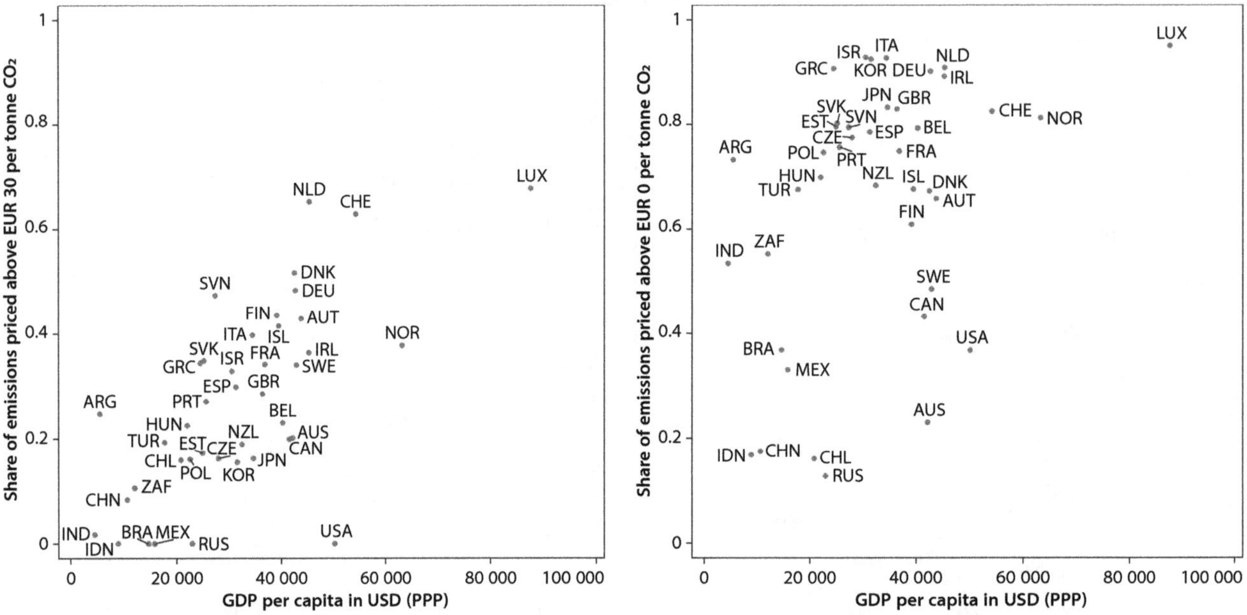

Source: GDP data is from the World Bank (2016), *World Development Indicators* (database), http://data.worldbank.org/data-catalog/world-development-indicators.

Countries' carbon pricing policies may be related to the share of energy imports in a country. Figure 4.19 shows the share of emissions in each country with ECR above EUR 30 per tonne of CO_2 (left-hand panel) and above EUR 0 (right-hand panel) against each country's net energy imports. Net energy imports are defined as a country's total energy use less its production, divided by its energy use. Energy-importing countries have a positive value for net energy imports, with 100% implying that the country imports all of its energy. Energy exporters show negative values for net energy imports. There is no lower limit for negative energy imports as countries can export more than their own energy use. One such country, Norway, is omitted from the graphs due to the very high proportion of energy exports (Norway exports 581% of its energy use, according to World Bank, 2016).

Net importers of energy tend to price a larger share of their emissions above EUR 0, as the right-hand panel of Figure 4.19 shows. This relationship is also observed with the share of emissions priced above EUR 30 per tonne of CO_2 (left-hand panel), although to a lesser extent.

To summarise, higher ECRs tend to be associated with higher and less carbon-intensive levels of GDP, and higher energy imports correlate with higher ECRs.

Figure 4.19. **Proportion of CO_2 emissions priced above EUR 30 (left) and EUR 0 (right) per tonne of CO_2 relative to net energy imports**

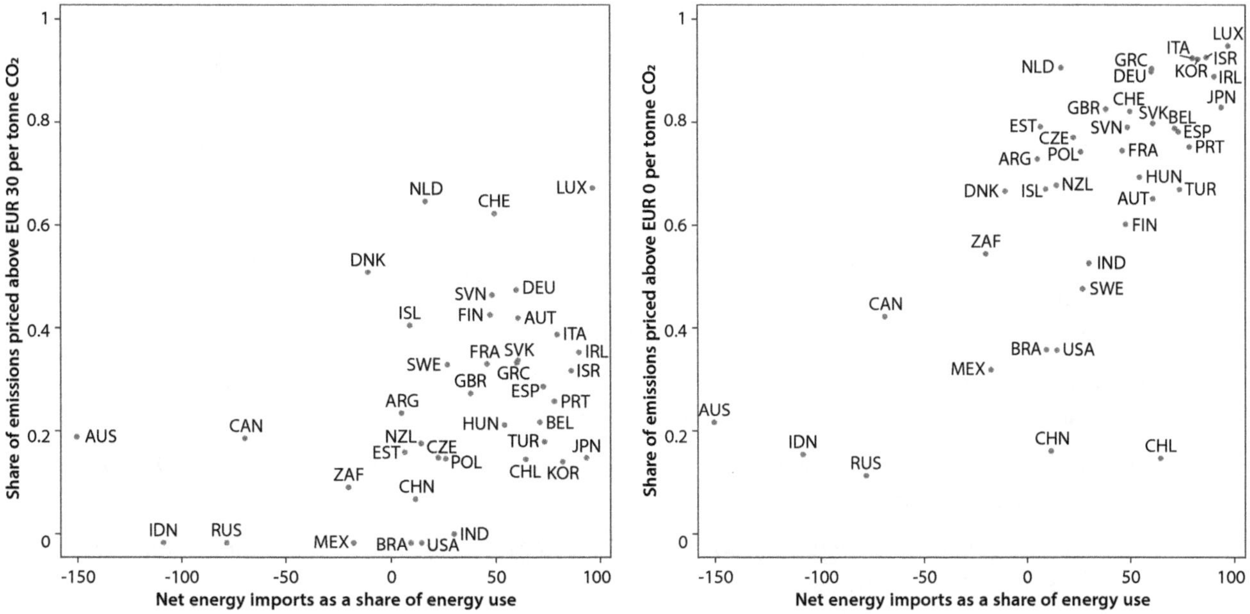

Source: Net energy imports are from the World Bank (2016), *World Development Indicators* (database), http://data.worldbank.org/data-catalog/world-development-indicators.

Notes

1. For example, if 50% of non-road emissions in a country face no carbon tax, 25% are subject to a carbon tax of EUR 10, and 25% to a carbon tax of EUR 20, then the average rate of the carbon tax component of the ECR would be EUR 7.5 (0.5 x 0 + 0.25 x 10 + 0.25 x 2 0).

2. The taxes on electricity consumption included in the *Taxing Energy Use* database are shown as indirect taxes on the fuels used to generate electricity, and are aggregated with specific taxes on energy on fuels used for the generation of electricity, where applicable. As explained in more detail in the TEU publications (OECD 2013 and OECD 2015a), the price signal from taxes on electricity consumption does not distinguish between the different fuels used for the generation of electricity, and thus does not encourage reducing the carbon intensity of fuel used to generate electricity.

3. Due to differences in data on emissions trading systems and taxes, there is some uncertainty associated to the proportion of emissions covered by both taxes and ETS in the industrial sector. Total combined ETS and tax coverage as presented in Figure 4.10 represent a point estimate of this overlap, which may be slightly higher or lower in reality. (For example, at a given ETS and tax coverage, the total coverage in the industry sector in the Czech Republic might be 10 percentage points higher or lower than indicated in Figure 4.10.) Annex A provides details on the methodology used to estimate this overlap.

4. In cases where taxes do not apply because a source is subject to an ETS, these taxes are not included in the tax rate shown in the figures.

References

European Commission (2016), "Structural reform of the European Carbon Market", http://ec.europa.eu/clima/policies/ets/reform/index_en.htm (accessed 8 March 2016).

Johansson, Å. et al. (2013), "Long-Term Growth Scenarios", *OECD Economics Department Working Papers*, No. 1000, OECD Publishing, Paris, http://dx.doi.org/10.1787/5k4ddxpr2fmr-en.

National Development and Reform Commission (2016), "National carbon emissions trading market to start", Notice [2016], No. 57, www.ndrc.gov.cn/zcfb/zcfbtz/201601/t20160122_772123.html (accessed 25 February 2016).

IEA (2016), "Extended world energy balances", *IEA World Energy Statistics and Balances* (database), International Energy Agency, Paris, http://dx.doi.org/10.1787/data-00513-en.

OECD (2016), "6A Value added and its components by activity, ISIC rev4", *OECD Stat*, (database), http://stats.oecd.org (accessed 11 Feb 2016).

OECD (2015), *Taxing Energy Use 2015: OECD and Selected Partner Economies*, OECD Publishing, Paris, http://dx.doi.org/10.1787/9789264232334-en.

OECD (2013), *Taxing Energy Use: A Graphical Analysis*, OECD Publishing, Paris, http://dx.doi.org/10.1787/9789264183933-en.

World Bank (2016), *World Development Indicators* (database), The World Bank, Washington, DC, http://data.worldbank.org/data-catalog/world-development-indicators.

Chapter 5

Effective carbon rates: Summary and conclusions

This chapter summarises the results of the analysis of effective carbon rates and draws some conclusions on the distribution of carbon prices across the economy and by sector, and on the composition of effective carbon rates.

Pricing carbon with market-based instruments is one of the most effective and lowest-cost means of inducing carbon abatement. While pricing alone is not sufficient to deliver the degree of abatement required for a low carbon transition, it is an essential part of the solution.

Two market-based mechanisms can be used to price carbon: taxes and emissions trading systems. Taxes can be further divided into those with a stated explicit purpose to induce CO_2 abatement (carbon taxes) and specific taxes on energy use, that also give rise to an effective tax rate on CO_2 emissions from energy use, even if levied for a mix of reasons. Whichever mechanism is used, pricing should cover emissions as broadly and uniformly as possible in order to ensure cost-effective abatement.

There are at least two approaches to valuing the climate cost of carbon emissions. The value can reflect the estimated marginal cost of climate change per unit of abatement, or it can be set as a function of the desired level of abatement. The marginal cost of climate change is very uncertain, with EUR 30 per tonne representing lower-end estimates in this report. There is considerable debate about appropriate levels of carbon pricing, a debate partially related to inter-country distributional concerns. The low-end estimate used here is not a recommendation on what prices to implement, but instead is merely a marker to gauge the current state of carbon pricing policies.

This report is the first combined and comprehensive analysis of the use of three carbon pricing instruments: specific taxes on energy use, carbon taxes, and emissions trading systems – together called the "effective carbon rate". It considers their levels and application across 41 OECD and G20 countries which together account for 80% of global CO_2 emissions from energy use. It finds that policy-induced prices of CO_2 emissions from energy use are often very low, even when the impact of all three pricing mechanisms is taken into account. Sixty percent of the emissions in these countries are not priced at all, either under emissions trading systems, carbon taxes, or specific taxes on energy use. Across the group of countries, only 10% of emissions are priced at or above EUR 30 per tonne of CO_2, and most of these emissions are from road transport energy use. Very low rates of carbon prices are found outside road transport, where 85% of total emissions originate. Within the non-road sectors, 70% of emissions face no carbon price at all, and only 4% face a price of more than EUR 30 per tonne of CO_2.

The low prices across the group of countries as a whole are in part due to the low prices or limited coverage by pricing mechanisms in several of the largest economies. Their weight in total emissions strongly influences the results. For example, the top five countries in terms of the quantity of emissions (China, the United States, India, Russia and Japan) together account for 70% of the emissions from all countries included in the analysis, but have comparatively low average ECRs on an economy-wide basis, with all of these countries but Japan featuring among the seven lowest economy-wide ECRs observed.

The high ECRs in road transport, relative to ECRs in other sectors, are observed both for the 41 countries as a whole, and also within every individual country. Within the road sector, almost all emissions are taxed at these comparatively high rates, including notably the emissions from gasoline and diesel use, which together comprise over 95% of all road emissions. However, in the road sector, it is very unusual for either carbon taxes or emissions trading systems to play a key role in pricing (notable exceptions include the carbon taxes in Finland, Norway, Sweden and British Columbia, and the emissions trading systems in California, Quebec and New Zealand). Consequently the price signal in the road sector is almost entirely derived from excise taxes.

As noted, however, road transport typically represents a small proportion of emissions, at 15% in total and varying between 6% in China to 38% in Slovenia (Iceland and Luxembourg are here excluded as outliers, in Iceland's case because of the extremely high share of renewable energy in electricity generation, which means that transport forms a disproportionately high share of emissions, and in the case of Luxembourg, because of the high volume of fuel sales for transit traffic). In all countries except Iceland and Luxembourg, non-road emissions represent more, and often significantly more, than 60% of total emissions; and 85% of total emissions across the 41 countries as a whole. It is in the non-road sector where much of the variation in pricing arises.

The non-road sectors of energy use include residential and commercial use, industry, electricity generation, offroad transport, and agriculture and fisheries. Of these sectors, the three biggest sources of emissions are industry, electricity generation and the residential and commercial sector (at 33.3%, 33.2% and 15.0%, of total CO_2 emissions from energy use, respectively). The other two sectors, agriculture and fisheries, and offroad transport account for only 1.2% and 2.2% of emissions respectively. Within countries, the relative weights of the sectors differ, but in every country, the former three sectors are considerably larger than the latter two.

The impact of the different pricing mechanisms varies considerably among these sectors and across the countries considered. In general, emissions in non-road sectors are subject to much lower tax rates than road transport, and in several cases, taxes are almost non-existent in these sectors. However, emissions trading systems apply most strongly outside road transport. Their coverage is largest in the industrial and electricity generation sectors. Emissions from residential and commercial energy use are more typically covered by taxes. Emissions trading systems do not affect prices or coverage substantially in this sector. Carbon taxes contribute very weakly to coverage and price levels in the non-road sectors across the 41 countries, almost entirely overlapping with taxes or ETS when they do apply. Within countries, they are most influential in Switzerland, Iceland, Ireland, Denmark, Finland, Norway, Slovenia and Sweden.

In the industrial sector, there is a degree of overlap in coverage between taxes and emissions trading systems, meaning that they act together to increase the carbon price beyond the level of either instrument alone – trading systems cover 13% of emissions, taxes apply to 17% of emissions, and total coverage is 26%. However, it is rare that taxes or emissions trading systems, whether applying separately or together, increase ECRs on industrial emissions beyond EUR 30 per tonne of CO_2. The overall impact on average ECRs in industry is highest from taxes when assessed across the 41 countries, but when isolating countries using an ETS, the two instruments are often more equal in price signals.

In the electricity sector, emissions trading systems and consumption taxes overlap almost entirely where trading systems exist. Consumption taxes on electricity fail to distinguish between the carbon-intensity of fuels used to generate electricity, but they do raise the price of emission-intensive electricity all the same.

The summary of results up to now defines emissions of energy as those from the combustion of energy, including combustion of biomass. Alternatively, biomass emissions can be excluded, for example if they are considered as carbon neutral on lifecycle emissions. Such alternative treatment does not affect the results across the 41 countries to any appreciable degree, but results for countries with large shares of biomass combustion are affected more strongly, particularly if those countries price non-biomass emissions highly. Biomass in road transport is relatively unimportant, but where it is used, ECRs are relatively high, reflecting relatively high rates on transport emissions in general. Emissions from non-road biomass combustion are usually exempt from or zero-rated in pricing mechanisms. Consequently, omitting these emissions from the ECR calculations increases coverage and average rates.

Summing up, the main features of the current landscape of carbon pricing instruments are as follows:

- Taxes are higher and more uniform in road transport, and lower and more variable in other sectors. They fairly consistently apply at highest rates to oil products. In electricity, taxes primarily relate to consumption. These patterns are observed in most countries.
- In countries where they exist, emissions trading systems typically have the highest coverage in the electricity and industrial sectors. Differences across countries in terms of implementation, coverage, rates and allowance allocation mechanisms are large. However, in all cases, allowance prices are uniform across the sources covered (in contrast to taxes) and outside road transport are more similar in magnitude to taxes.
- Carbon tax coverage is low in most of the countries where carbon taxes apply. Carbon taxes almost always apply in countries that also implement an emissions trading system.
- The combined impact of low taxes and the low and still limited deployment of emissions trading systems outside road transport leads to the conclusion that most emissions are not priced at all or to a low degree.

While this report did not aim to investigate how these main features have come about, some observations were made. For example, more emissions are priced in countries with higher per capita GDP and with a lower carbon-intensity of GDP. Causation can run in two directions, with high rates having induced more abatement or low carbon dependence allowing higher ECRs. Also, countries that import a large share of their energy use are more likely to price CO_2 emissions from energy use, possibly related to energy security or trade balance concerns.

The principal contribution of this report is to present the distribution of ECRs on energy use. These distributions describe what share of emissions is subject to at least (or at most) a given level of carbon price. In addition, these shares are attributed to sectors, and the composition of the ECRs is presented. A summary measure of the distribution is introduced, namely the "carbon pricing gap".

The carbon pricing gap measures the extent by which carbon pricing falls short of the EUR 30 benchmark, as a percentage of total emissions. For current ECRs, the carbon pricing gap is 80.1%, meaning that less than 20% of emissions are priced at EUR 30 per tonne of CO_2 or more. If prices and coverage in all countries were to increase to at least the levels currently observed at the median for each sector, the carbon pricing gap would decline to 53.1%, suggesting that significant progress can be made by increasing rates where they are currently low, or implementing them where no carbon price applies currently. Such an approach would also lead to more uniform carbon prices, and this would result in more cost-effective abatement. Cost-effectiveness is always a criterion worth considering, and it becomes more important as abatement targets become more stringent.

Part II

Country results

Chapter 6

Effective carbon rates across 41 countries and on a country-by-country basis

This chapter presents the results of the analysis of effective carbon rates for the 41 countries individually. For each country, the chapter provides full detail of the distribution of effective carbon rates across carbon emissions from energy use. In addition, the chapter depicts the composition of effective carbon rates by price instrument (taxes and price signals from emissions trading systems) for six economic sectors in each of the 41 countries.

This part of the report presents the results of the analysis of effective carbon rates (ECRs) on a country-by-country basis. In particular, it provides full details of the proportion of CO_2 emissions from energy use subject to different ECR levels in each of the 41 countries, as well as the level and components of average ECRs in each of the six economic sectors (road transport, offroad transport, industry, agriculture and fisheries, residential and commercial, and electricity). To facilitate the interpretation of the graphs on a country-by-country basis, the figures are first presented and explained for the group of 41 countries as a whole.

Distribution of effective carbon rate levels

The proportion of CO_2 emissions from energy use subject to different ECR levels, and the composition of ECRs by sector, are analysed in Part I of this report. The discussion here provides intuition for the interpretation of the country-by-country graphs presented in this part.

Figure 6.1 presents the full distribution of ECR levels across the 41 economies, sorting CO_2 emissions from energy use according to the ECR at which they are priced, starting at zero. The horizontal axis shows the proportion of CO_2 emissions while the vertical axis shows the ECR in EUR per tonne of CO_2. Figure 6.1 allows the reader to identify the share of CO_2 emissions from energy use priced at any given ECR. For example, the first dot in Figure 6.1 highlights that 90% of carbon emissions from energy use across the 41 countries are priced below an effective carbon rate of EUR 30, and other dots along the ECR series highlight the share of carbon emissions priced at or below other levels of ECRs (e.g. EUR 50, EUR 100 or EUR 120 per tonne of CO_2). The use of particular values does not imply a normative statement about a minimum level of carbon pricing that should be implemented. Country-specific versions of Figure 6.1 are used in this part of this report to show the pattern of ECRs in each of the 41 countries.

Figure 6.1. **Proportion of CO_2 emissions from energy use subject to different levels of effective carbon rates across 41 OECD and selected partner economies**

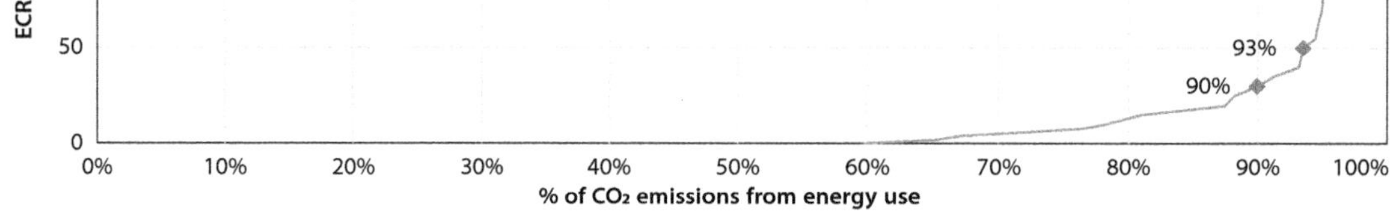

StatLink http://dx.doi.org/10.1787/888933390210

Average effective carbon rates by sector and price instrument

The levels and coverage of effective carbon rates vary widely across different sectors of the economy. This is illustrated in Figure 6.2, which presents the average ECRs and carbon emissions from energy use for six sectors of the economy (road transport, offroad transport, industry, agriculture and fisheries, residential and commercial, and electricity). The horizontal axis of Figure 6.2 shows total CO_2 emissions from energy use for each sector in the 41 countries in thousand tonnes of CO_2. The width of each sector along the horizontal axis therefore represents the total carbon emissions from energy use from each sector, which is also displayed in Table 4.3 in Part I. The vertical axis shows different levels of ECRs. Within each of the six sectors, the width of the shaded rectangles shows the amount of CO_2 emissions from energy use in that sector subject to each type of price instrument. The height of each shaded rectangle represents the average price signal from that instruments for all emissions priced by that instrument (i.e. zeros are excluded).

Figure 6.2. **Average effective carbon rates across the 41 economies by sector and component**

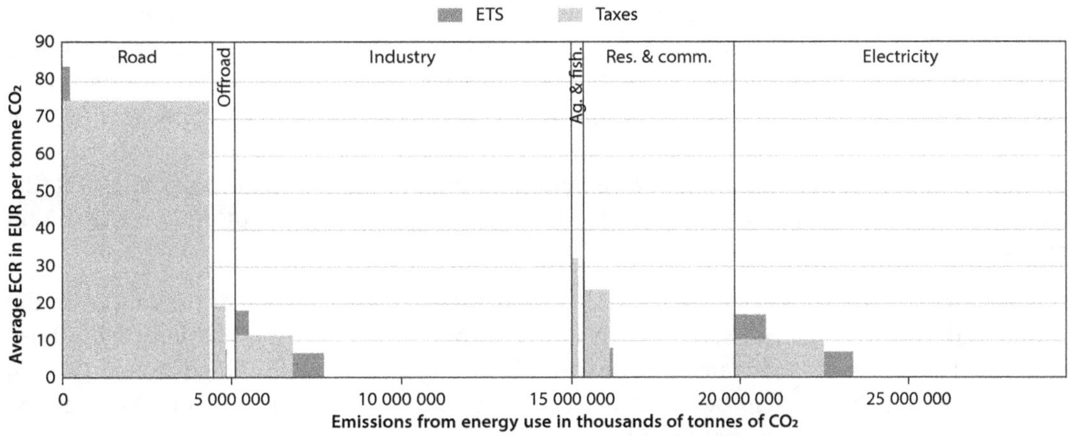

StatLink http://dx.doi.org/10.1787/888933390631

Figure 6.2 allows the components of the average ECR in each sector to be identified. Carbon taxes and specific taxes from energy use are shown in light blue, while the ETS components of average ECRs are shown in dark blue. A tonne of CO_2 emissions can be priced in different ways: only via taxes (a light blue rectangle), only via a tradable emissions permit price (a dark blue rectangle), or via taxes and tradable permits (a light blue rectangle, with a dark blue rectangle on top). Emissions that are not priced are shown in the base, with no light blue or dark blue rectangle.

Figure 6.2 illustrates the wide variation in the composition of average ECRs in the different sectors of the economy across the 41 countries. Ninety-eight percent of carbon emissions from road transport are priced (shown as the width of the light blue rectangle in the road transport sector), at an average of EUR 75 per tonne of CO_2. ECRs in road transport are almost entirely constituted by taxes, and only 5% of road emissions are subject to an ETS (the narrow dark blue bar). Where an ETS applies in the road sector, the combined average price signal for the emissions covered by the ETS from taxes and the ETS is EUR 84 per tonne of CO_2. CO_2 emissions from offroad transport and from agriculture and fisheries – two relatively small sectors in terms of CO_2 emissions – are mainly priced via taxes, but ECR coverage is significantly lower than in the road sector (see the width of the light blue bars), and they are also priced at lower average rates (see the height of the light blue bars). ETS only cover a very small proportion of emissions in each of these sectors, as shown by the almost invisible dark blue bar in these two sectors.

Emissions from industry account for a third of carbon emissions from energy use across the 41 countries. Seventy-four percent of industrial emissions are unpriced. Of the remainder, the instrument mix is relatively diverse: 13% of emissions are subject to taxes only (the light blue bar), 9% of emissions are priced via an ETS only (the dark blue bar), and roughly 4% of emissions are priced via both taxes and ETS (the dark blue and light blue bar combined). The overlap between ETS and taxes is thus relatively small in the industry sector, implying that the two pricing instruments often cover different emissions, increasing the total amount of emissions covered by a price. The average ECR is highest when taxes and ETS overlap (EUR 18.13 per tonne of CO_2 on average), while carbon prices from tradable emissions permits send the lowest average price signal in this sector (EUR 6.6 per tonne of CO_2 on average).

A similar pricing pattern is observed for carbon emissions from energy used in electricity generation, which also account for roughly a third of emissions in the 41 countries on average. While in this sector 64% of carbon emissions are unpriced, 9.5% of emissions are subject to both a tax and an ETS and which face an average combined price from these instruments of EUR 17 per tonne CO_2 (i.e. the light blue and dark blue bar). Taxes on energy use (the light blue bar in the electricity sector) apply to 17.4% of emissions at a rate of EUR 10.19 per tonne of CO_2 on average. A lower proportion of emissions (8.6%) are priced at EUR 6.18 per tonne of CO_2 on average through an ETS (the dark blue bar). As in the industry sector, the taxes and permit prices often cover different emissions.

The price signal from energy taxes dominates the ECR on carbon emissions from the residential and commercial sector, as it covers a relatively large proportion of the base (16.7%). The average ECR rate from tax and ETS together is EUR 31 per tonne of CO_2 on average, but the overlap between the two instruments is restricted to a very narrow proportion of the emissions base (1%).

Because different fuels used in one sector may be taxed at very different rates, the average carbon rates presented in Figure 6.2 may hide significant differences within sectors. For example, a majority of countries tax CO_2 emissions from gasoline used in the road transport sector at much higher rates than those from diesel. Similarly, emissions from coal use are often taxed at rates significantly lower than those applied to emissions from oil products or natural gas. *Taxing Energy Use* (OECD, 2015, 2013) provides more detail on the exact distribution of tax rates that underlies the average tax rates shown in Figure 6.2 and the country versions of this figure presented below.

All remaining figures in this part of the report include emissions from biomass. Where emissions from biomass account for a large share of a country's carbon emissions, figures excluding emissions from biomass (i.e. taking a lifecycle approach to accounting for the emissions from biomass), are cited in the accompanying text.

Annex A describes the detailed methodology and the data sources used for calculating ECRs, while Annex B provides additional information on estimating coverage and permit prices of emissions trading systems for calculating ECRs.

Argentina

Figure 6.3. **Proportion of CO_2 emissions from energy use subject to different levels of effective carbon rates in Argentina in 2012**

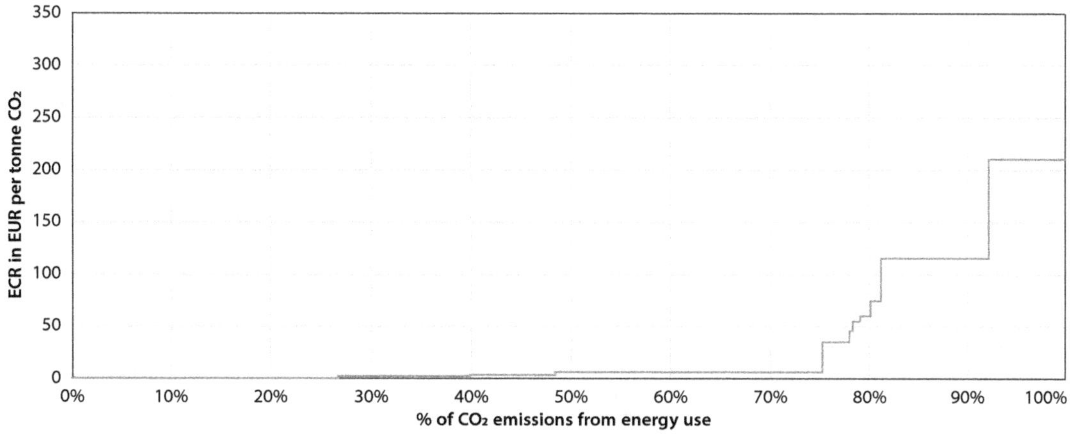

StatLink http://dx.doi.org/10.1787/888933390225

Figure 6.4. **Average effective carbon rates in Argentina by sector and component in 2012**

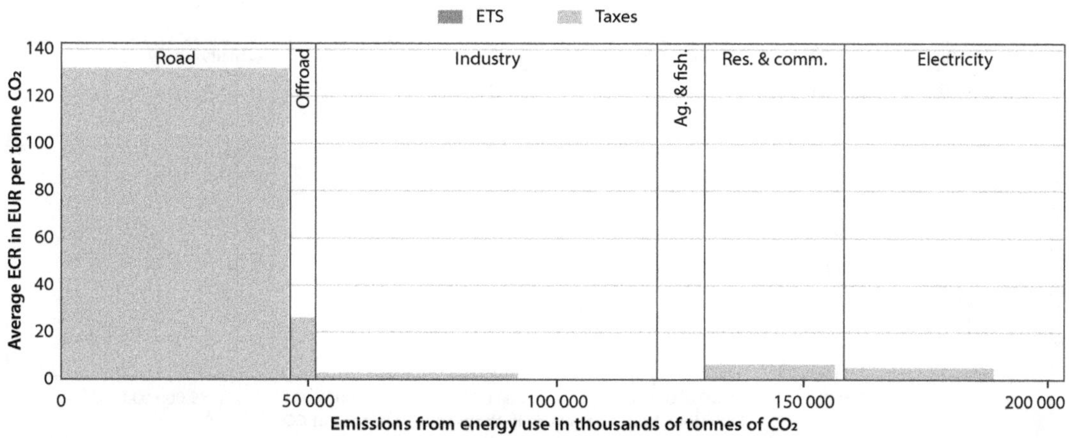

StatLink http://dx.doi.org/10.1787/888933390643

Note: See Box 3.1 in Chapter 3 for notes on the interpretation of effective carbon rates. Please see Annex A and Annex B for detail on methodology and underlying sources.

In 2012, Argentina priced 73% of its energy related CO_2 emissions through specific taxes on energy use. It did not apply explicit carbon taxes or emissions trading systems. A price above EUR 30 per tonne of CO_2 applied to 25% of emissions (see Figure 6.3), all of which were from the transport sector. Carbon emissions from energy use in agriculture and fisheries were not priced. The majority of unpriced emissions were from energy use in the industry and electricity sectors (see Figure 6.4).

Australia

Figure 6.5. **Proportion of CO_2 emissions from energy use subject to different levels of effective carbon rates in Australia in 2012**

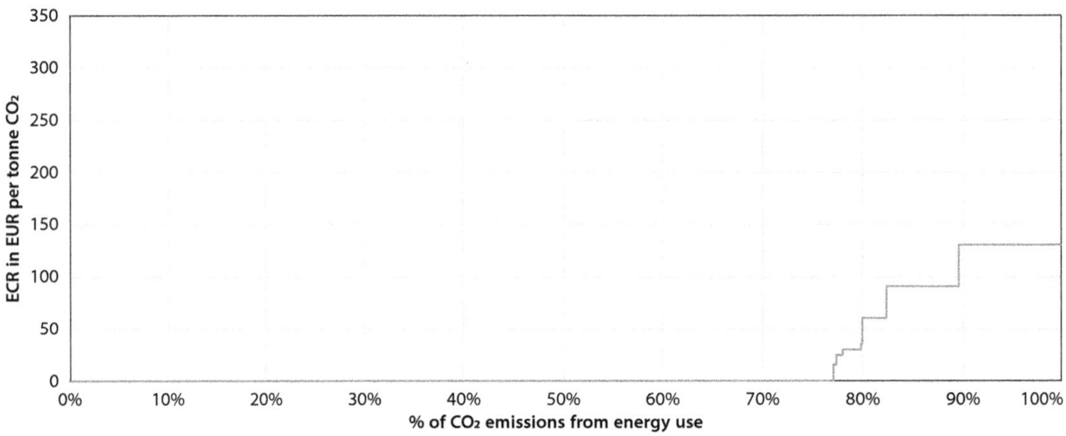

StatLink http://dx.doi.org/10.1787/888933390231

Figure 6.6. **Average effective carbon rates in Australia by sector and component in 2012**

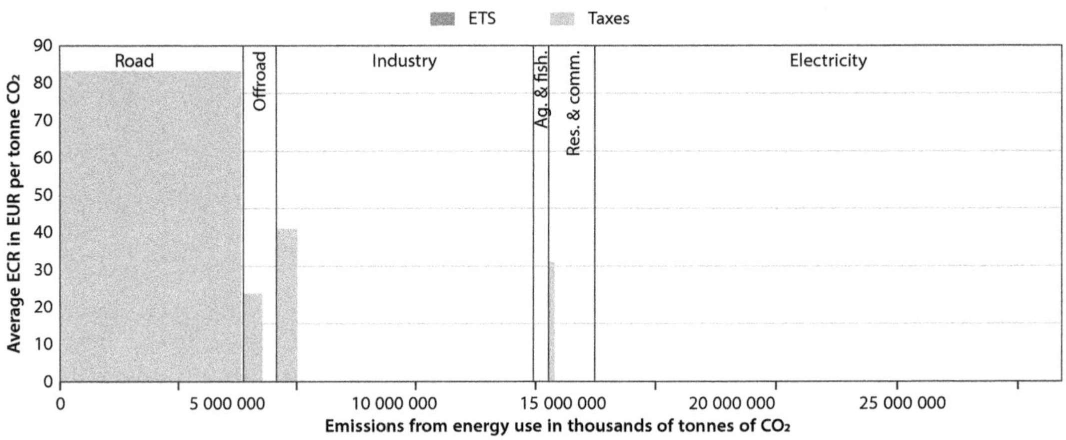

StatLink http://dx.doi.org/10.1787/888933390650

Note: See Box 3.1 in Chapter 3 for notes on the interpretation of effective carbon rates. Please see Annex A and Annex B for detail on methodology and underlying sources.

In 2012, effective carbon rates in Australia arose from specific taxes on energy use. Australia did not apply an explicit carbon tax or an emissions trading system. Australia priced 23% of carbon emissions from energy use. An ECR above EUR 30 per tonne of CO_2 applied to 20% of emissions, the majority of which was from road transport (see Figure 6.5). Carbon emissions from energy use in electricity and agriculture and fisheries were not priced. The majority of unpriced emissions in Australia stemmed from energy used in the industry and the electricity sectors (Figure 6.6).

Austria

Figure 6.7. **Proportion of CO_2 emissions from energy use subject to different levels of effective carbon rates in Austria in 2012**

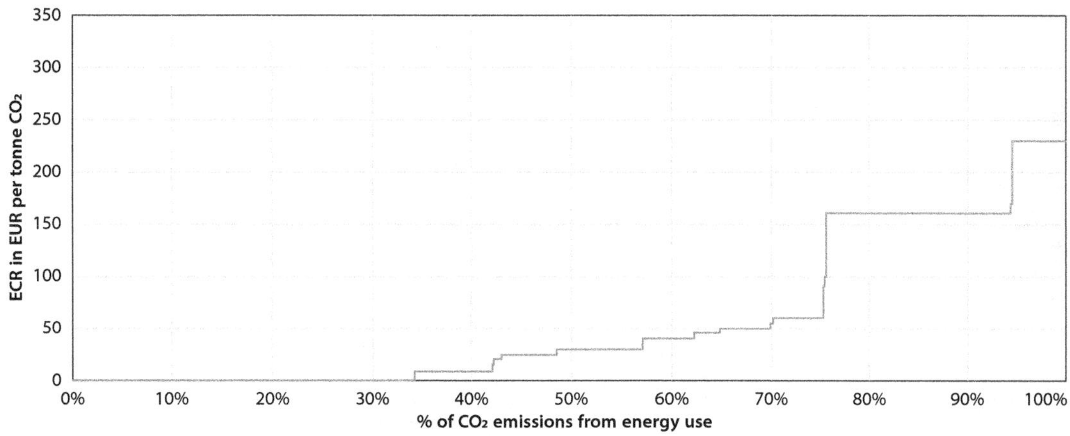

StatLink http://dx.doi.org/10.1787/888933390247

Figure 6.8. **Average effective carbon rates in Austria by sector and component in 2012**

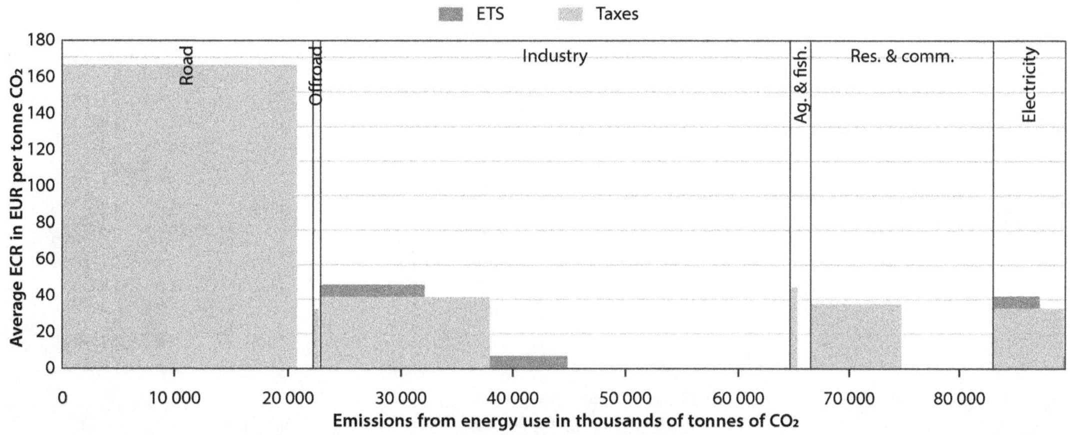

StatLink http://dx.doi.org/10.1787/888933390667

Note: See Box 3.1 in Chapter 3 for notes on the interpretation of effective carbon rates. Please see Annex A and Annex B for detail on methodology and underlying sources.

In 2012, effective carbon rates in Austria consisted primarily of specific taxes on energy use, and to a small extent of permit prices from the EU ETS. Austria did not have an explicit carbon tax. Austria priced 66% of its energy related CO_2 emissions and 43% were priced above EUR 30 per tonne of CO_2: a large share of the latter was from the energy used in road transport (see Figure 6.7). Unpriced emissions were primarily emitted by industry and the residential and commercial sector. The EU ETS alone covered 16.5% of carbon emissions from industry (see Figure 6.8).

Belgium

Figure 6.9. **Proportion of CO_2 emissions from energy use subject to different levels of effective carbon rates in Belgium in 2012**

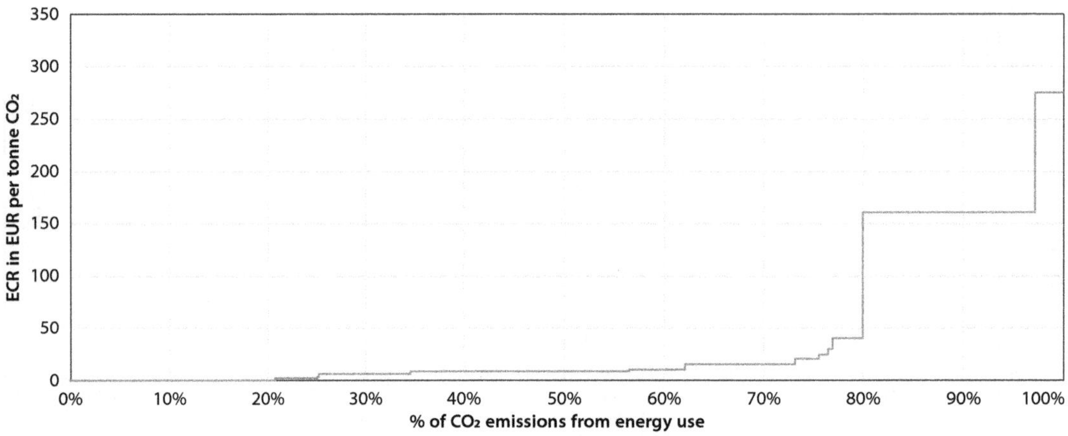

StatLink http://dx.doi.org/10.1787/888933390254

Figure 6.10. **Average effective carbon rates in Belgium by sector and component in 2012**

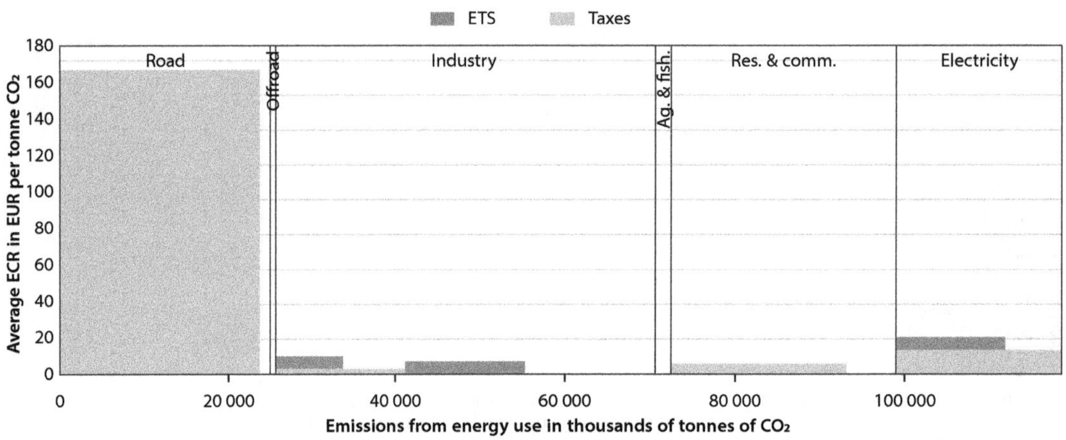

StatLink http://dx.doi.org/10.1787/888933390670

Note: See Box 3.1 in Chapter 3 for notes on the interpretation of effective carbon rates. Please see Annex A and Annex B for detail on methodology and underlying sources.

In 2012, effective carbon rates in Belgium consisted primarily of specific taxes on energy use and a smaller part of the ECR was from permit prices from the EU ETS. Belgium did not apply an explicit carbon tax. Belgium priced 79% of carbon emissions from energy use, and 23% were priced above EUR 30 per tonne of CO_2 (see Figure 6.9). The majority of these emissions were from the road sector. Carbon emissions from energy use in agriculture and fisheries were entirely unpriced. Unpriced emissions were primarily emitted by industry but also by the residential and commercial sector (see Figure 6.10).

Brazil

Figure 6.11. **Proportion of CO_2 emissions from energy use subject to different levels of effective carbon rates in Brazil in 2012**

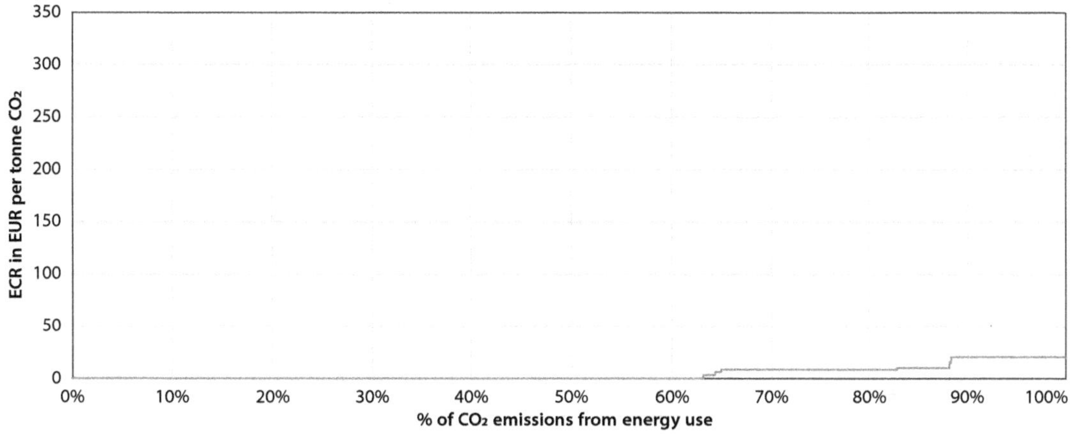

StatLink http://dx.doi.org/10.1787/888933390262

Figure 6.12. **Average effective carbon rates in Brazil by sector and component in 2012**

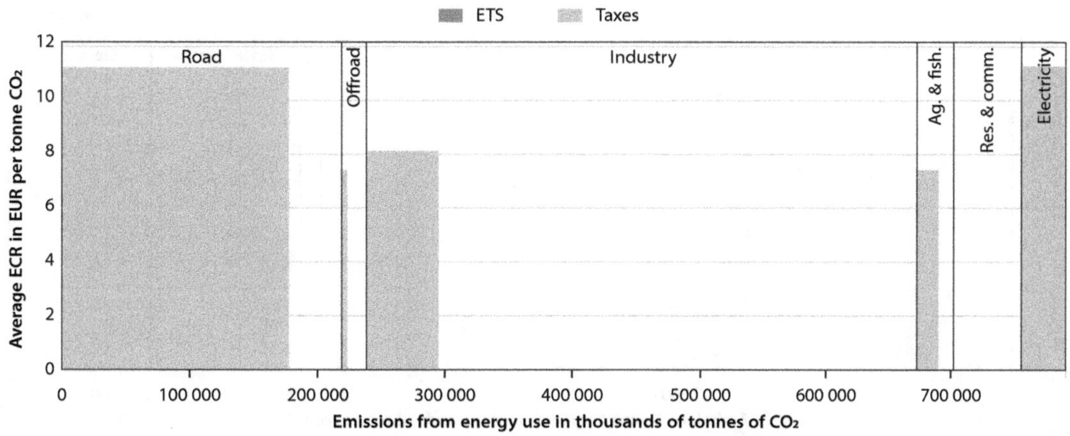

StatLink http://dx.doi.org/10.1787/888933390688

Note: See Box 3.1 in Chapter 3 for notes on the interpretation of effective carbon rates. Please see Annex A and Annex B for detail on methodology and underlying sources.

In 2012, effective carbon rates in Brazil consisted entirely of specific taxes on energy use. Brazil did not apply an explicit carbon tax or an emissions trading system. Brazil priced 37% of carbon emissions from energy use: no emissions were priced above EUR 30 per tonne of CO_2 (see Figure 6.11). The majority of priced emissions were from road transport. Carbon emissions from energy use in the residential and commercial sector were not priced (see Figure 6.12). The majority of unpriced emissions were from the industry sector.

Canada

Figure 6.13. **Proportion of CO_2 emissions from energy use subject to different levels of effective carbon rates in Canada in 2012**

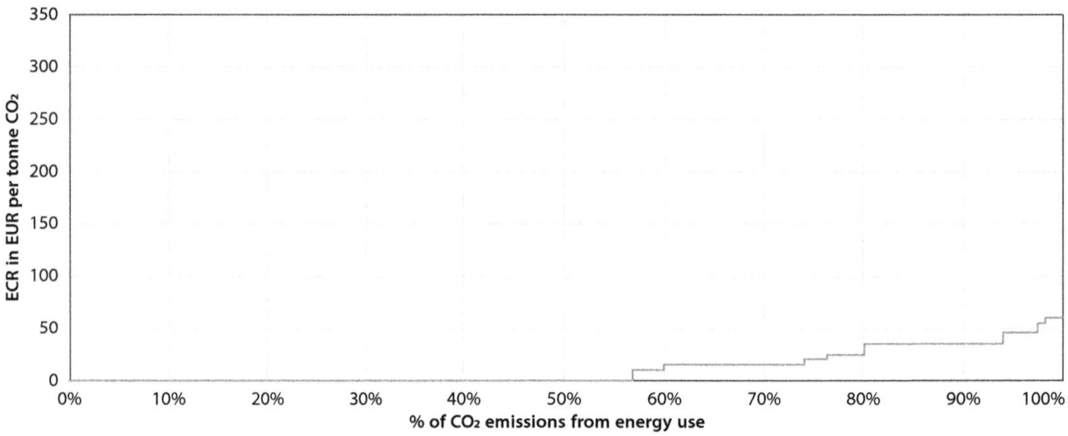

StatLink http://dx.doi.org/10.1787/888933390276

Figure 6.14. **Average effective carbon rates in Canada by sector and component in 2012**

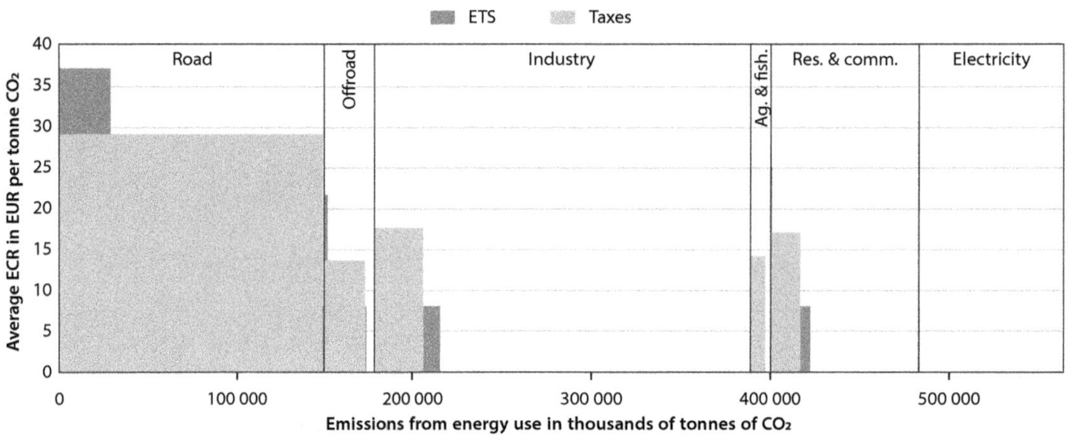

StatLink http://dx.doi.org/10.1787/888933390690

Note: See Box 3.1 in Chapter 3 for notes on the interpretation of effective carbon rates. Please see Annex A and Annex B for detail on methodology and underlying sources.

In 2012, effective carbon rates in Canada consisted primarily of specific federal taxes on energy use. Other carbon pricing instruments applied only at subnational levels: a carbon tax in British Columbia and an ETS in Quebec.[1] Canada priced 43% of its energy related CO_2 emissions, and 20% were priced above EUR 30 per tonne of CO_2 (see Figure 6.13). The majority of the priced emissions were from the road transport sector. The majority of unpriced emissions were emitted by industry and the electricity sector. In total, 9% of Canadian carbon emissions from energy use were covered by the Quebec ETS. The overlap between emissions covered by the Quebec Cap-and-Trade Program and federal taxes was very small (see Figure 6.14).

Chile

Figure 6.15. **Proportion of CO_2 emissions from energy use subject to different levels of effective carbon rates in Chile in 2012**

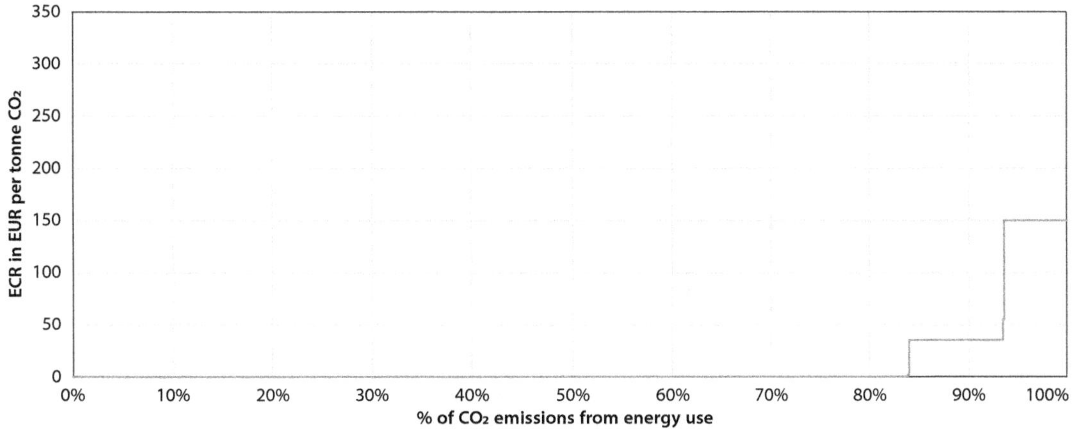

StatLink http://dx.doi.org/10.1787/888933390288

Figure 6.16. **Average effective carbon rates in Chile by sector and component in 2012**

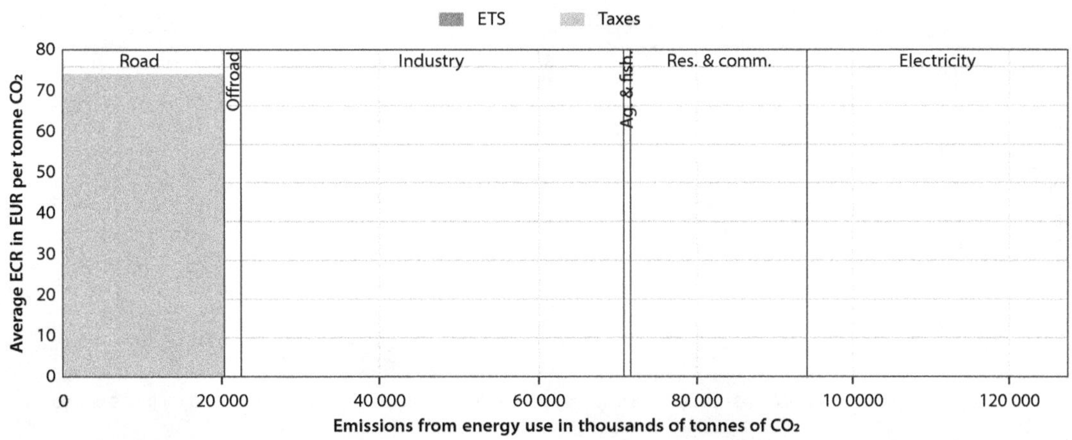

StatLink http://dx.doi.org/10.1787/888933390709

Note: Note: See Box 3.1 in Chapter 3 for notes on the interpretation of effective carbon rates. Please see Annex A and Annex B for detail on methodology and underlying sources.

In 2012, effective carbon rates in Chile consisted entirely of specific taxes on energy use. Chile did not apply an explicit carbon tax or an emissions trading system in 2012. Chile priced 16% of its energy related CO_2 emissions. All of these were priced at EUR 30 per tonne of CO_2 (see Figure 6.15), and were emitted by road transport (Figure 6.16). A carbon tax is scheduled for introduction in 2018, which will increase the share of emissions subject to a price.

People's Republic of China

Figure 6.17. **Proportion of CO_2 emissions from energy use subject to different levels of effective carbon rates in China in 2012**

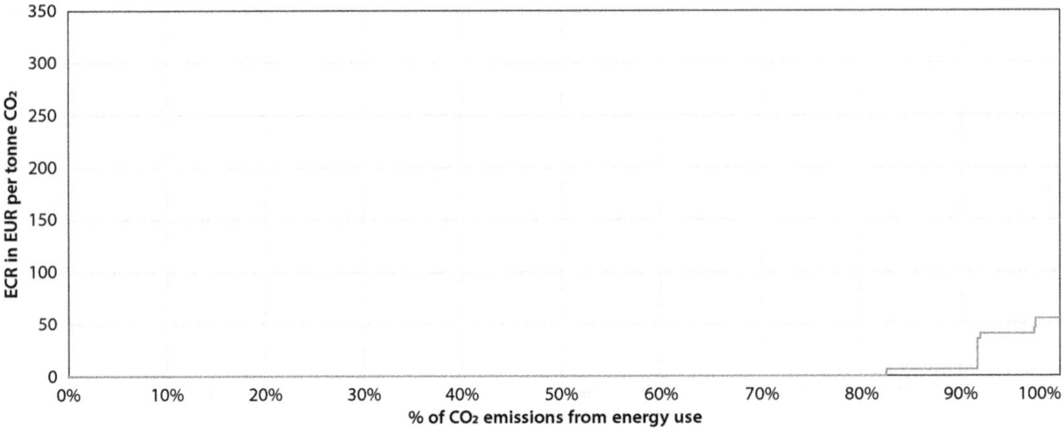

StatLink http://dx.doi.org/10.1787/888933390292

Figure 6.18. **Average effective carbon rates in China by sector and component in 2012**

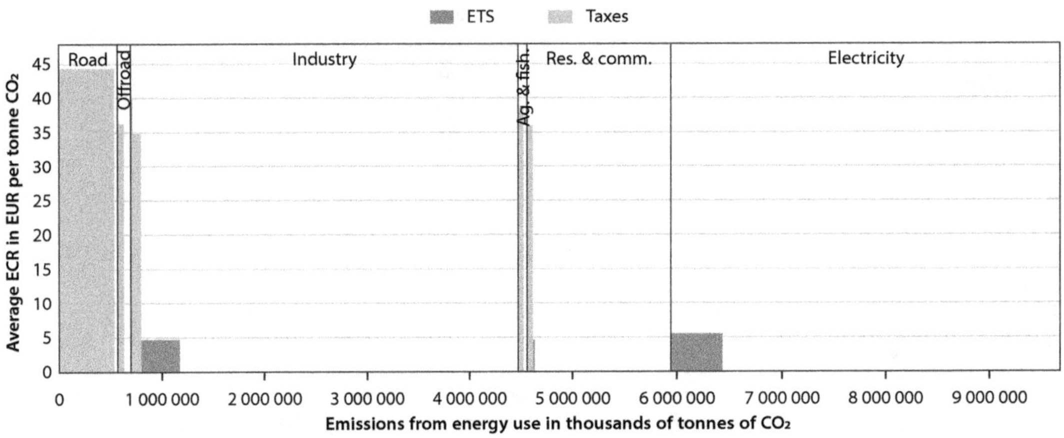

StatLink http://dx.doi.org/10.1787/888933390716

Note: Note: See Box 3.1 in Chapter 3 for notes on the interpretation of effective carbon rates. Please see Annex A and Annex B for detail on methodology and underlying sources.

In 2012, effective carbon rates in China consisted primarily of specific taxes on energy use. Five municipalities, Beijing, Chongqing, Shanghai, Shenzhen and Tianjin, and two provinces, Guangdong and Hubei have implemented emissions trading systems.[2] Eighteen percent of carbon emissions from energy use in China were covered by a price, and 8% were priced above EUR 30 per tonne of CO_2 (see Figure 6.17). The majority of these emissions stemmed from the road sector. Unpriced emissions were found primarily in the industry, the residential and commercial, and the electricity sectors (see Figure 6.18). In total, 9% of Chinese emissions were estimated to be covered by the subnational emissions trading systems. The overlap between the emissions covered by taxes and emissions trading systems was very small.

Czech Republic

Figure 6.19. **Proportion of CO_2 emissions from energy use subject to different levels of effective carbon rates in the Czech Republic in 2012**

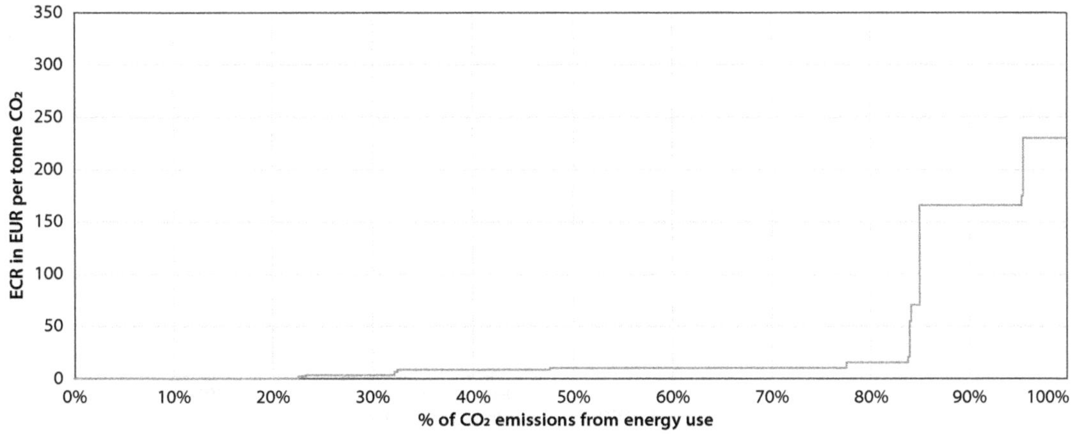

StatLink http://dx.doi.org/10.1787/888933390306

Figure 6.20. **Average effective carbon rates the Czech Republic by sector and component in 2012**

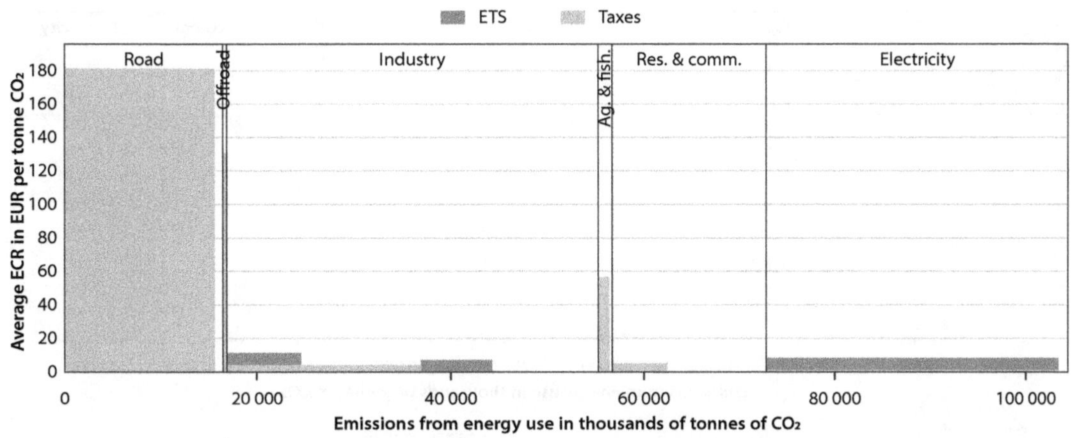

StatLink http://dx.doi.org/10.1787/888933390724

Note: Note: See Box 3.1 in Chapter 3 for notes on the interpretation of effective carbon rates. Please see Annex A and Annex B for detail on methodology and underlying sources.

In 2012, effective carbon rates in the Czech Republic consisted primarily of specific taxes on energy use and to a lesser extent of permit prices from the EU ETS. The Czech Republic did not have an explicit carbon tax. The Czech Republic priced 77% of its energy related CO_2 emissions, and 16% were priced at an ECR above EUR 30 per tonne of CO_2 (see Figure 6.19). Emissions priced at this level were primarily emitted by road transport (see Figure 6.20). More than 30% of emissions from the industry sector were covered by taxes, roughly 20% was covered by the EU ETS, and the coverage of taxes and ETS overlapped for another 20% of emissions in that sector. Permit prices from the EU ETS were a significant component of the ECR in the electricity sector.

Denmark

Figure 6.21. **Proportion of CO_2 emissions from energy use subject to different levels of effective carbon rates in Denmark in 2012**

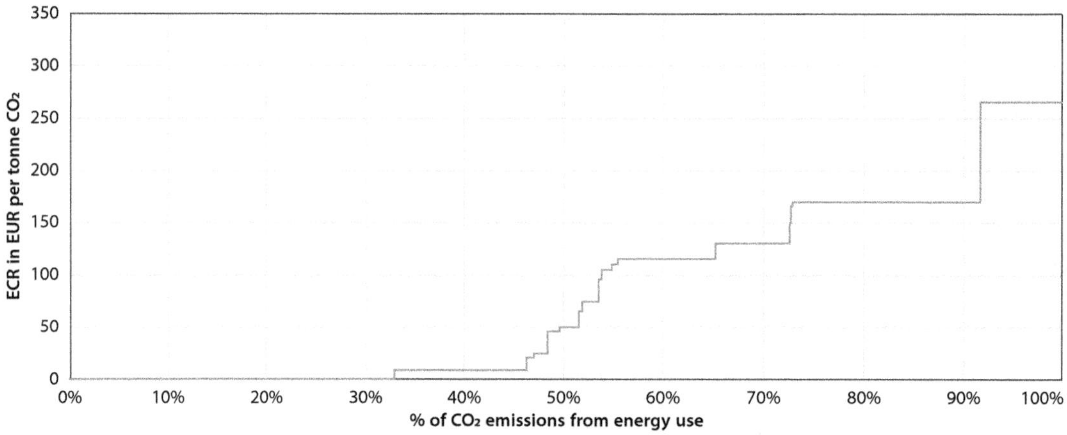

StatLink http://dx.doi.org/10.1787/888933390312

Figure 6.22. **Average effective carbon rates in Denmark by sector and component in 2012**

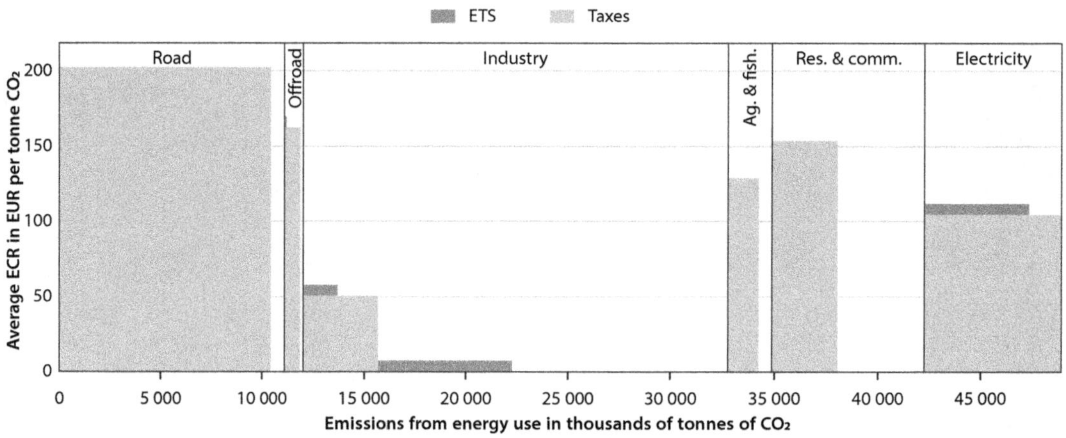

StatLink http://dx.doi.org/10.1787/888933390731

Note: Note: See Box 3.1 in Chapter 3 for notes on the interpretation of effective carbon rates. Please see Annex A and Annex B for detail on methodology and underlying sources.

In 2012, effective carbon rates in Denmark consisted primarily of specific taxes on energy use and to a lesser extent of national carbon taxes and permit prices from the EU ETS. Denmark priced 67% of its energy related CO_2 emissions, and 52% were priced at an ECR above EUR 30 per tonne of CO_2 (see Figure 6.21). Many of these emissions were from the road sector. The majority of unpriced emissions were from the industry and the residential and commercial sectors. Taxes and the EU ETS applied to a largely separate emissions base (see Figure 6.22).

A large share of unpriced emissions was from the combustion of biomass. Excluding emissions from biomass combustion, 92% of CO_2 emissions from energy use in Denmark were priced at a positive rate, and 69% were priced above EUR 30 per tonne of CO_2.

Similarly, when excluding biomass, 99% of total CO_2 emissions were priced at an ECR above EUR 30 per tonne of CO_2 in the residential and commercial sector. In the industry sector, 81% of emissions were priced, and 18% were priced above EUR 30 per tonne of CO_2 when excluding biomass.

Estonia

Figure 6.23. **Proportion of CO_2 emissions from energy use subject to different levels of effective carbon rates in Estonia in 2012**

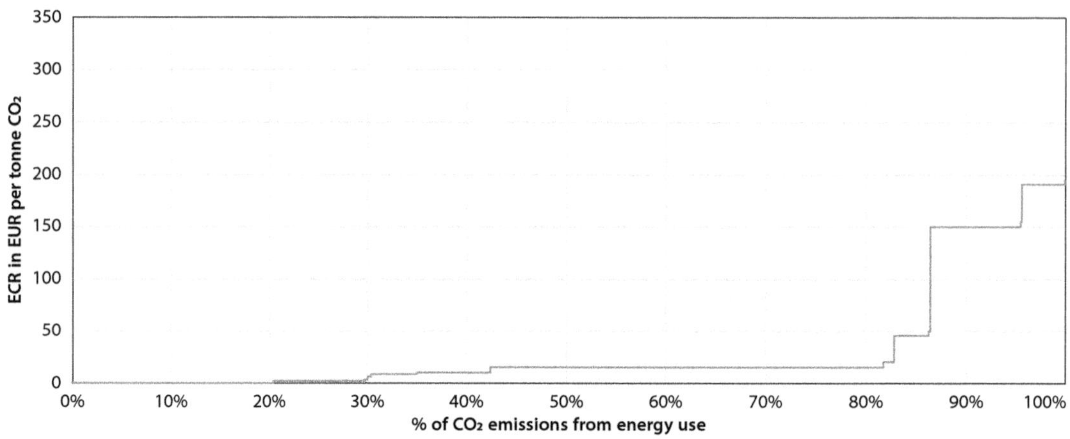

StatLink http://dx.doi.org/10.1787/888933390320

Figure 6.24. **Average effective carbon rates in Estonia by sector and component in 2012**

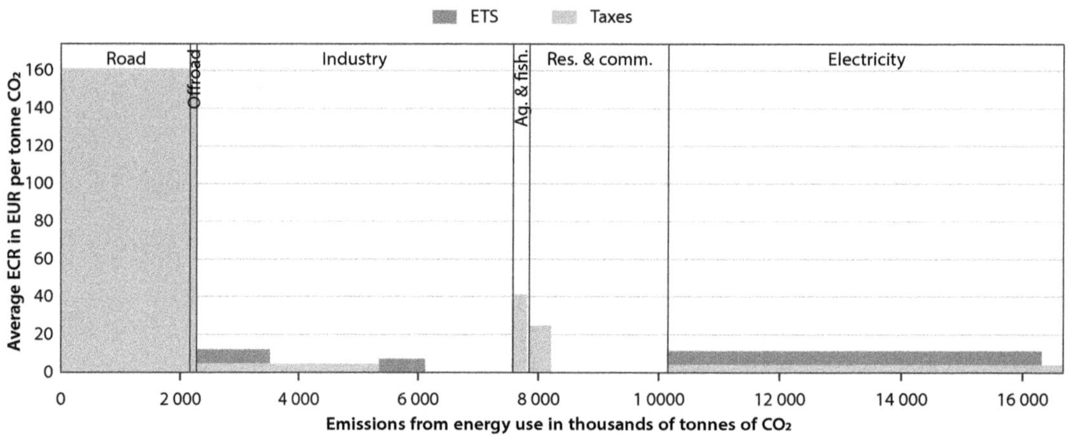

StatLink http://dx.doi.org/10.1787/888933390741

Note: Note: See Box 3.1 in Chapter 3 for notes on the interpretation of effective carbon rates. Please see Annex A and Annex B for detail on methodology and underlying sources.

In 2012, effective carbon rates in Estonia consisted primarily of specific taxes on energy use and to a lesser extent of permit prices from the EU ETS. Estonia priced 80% of its CO_2 emissions from energy use, and 17% were priced above EUR 30 per tonne of CO_2 (see Figure 6.23); the majority of these emissions was from road transport. Unpriced emissions were primarily emitted by industry, and the residential and commercial sectors (see Figure 6.24). Permit prices from the EU ETS accounted for a significant component of the ECR in the electricity sector.

Finland

Figure 6.25. **Proportion of CO_2 emissions from energy use subject to different levels of effective carbon rates in Finland in 2012**

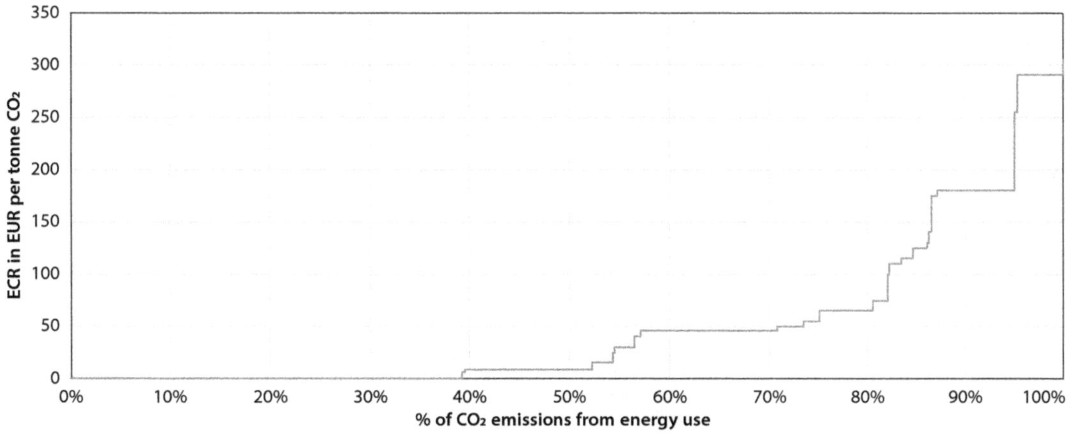

StatLink http://dx.doi.org/10.1787/888933390333

Figure 6.26. **Average effective carbon rates in Finland by sector and component in 2012**

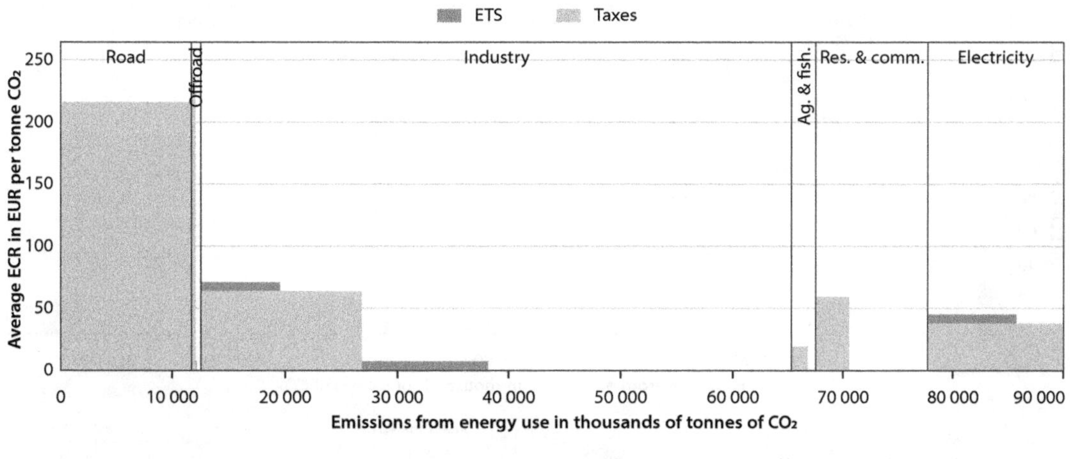

StatLink http://dx.doi.org/10.1787/888933390756

Note: Note: See Box 3.1 in Chapter 3 for notes on the interpretation of effective carbon rates. Please see Annex A and Annex B for detail on methodology and underlying sources.

In 2012, effective carbon rates in Finland consisted primarily of specific taxes on energy use and to a lesser extent of national carbon taxes and permit prices from the EU ETS. Finland priced 61% of carbon emissions from energy use, and 44% were priced above EUR 30 per tonne of CO_2 (see Figure 6.25); a significant share of these emissions was from road transport. Looking at the emissions from the industry sector that are subject to a price, the coverage of taxes and prices from tradable emissions permits overlapped by 27% of industrial emissions. The majority of unpriced emissions were emitted by industry and the residential and commercial sector (see Figure 6.26).

France

Figure 6.27. **Proportion of CO_2 emissions from energy use subject to different levels of effective carbon rates in France in 2012**

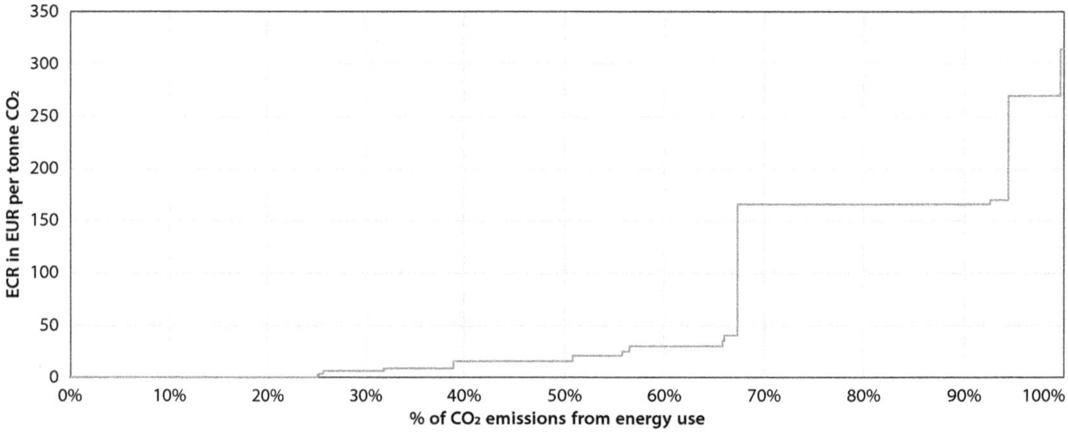

StatLink http://dx.doi.org/10.1787/888933390343

Figure 6.28. **Average effective carbon rates in France by sector and component in 2012**

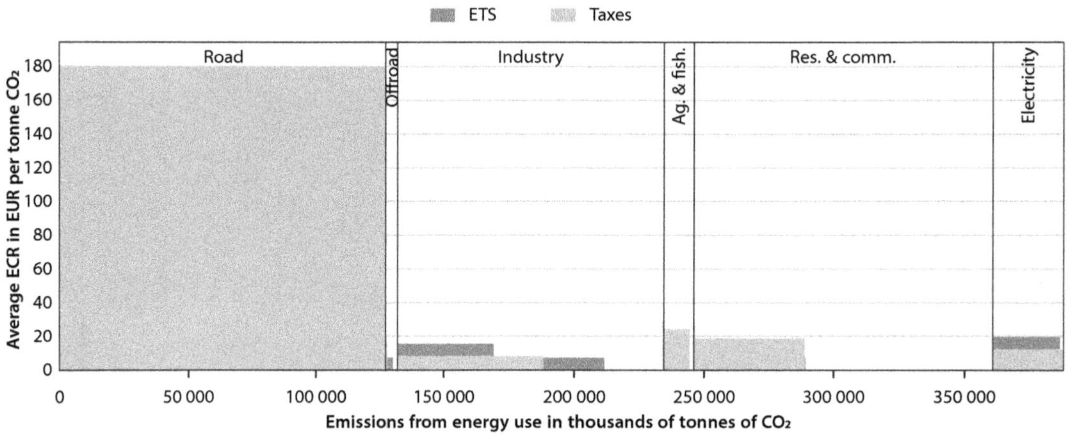

StatLink http://dx.doi.org/10.1787/888933390762

Note: Note: See Box 3.1 in Chapter 3 for notes on the interpretation of effective carbon rates. Please see Annex A and Annex B for detail on methodology and underlying sources.

In 2012, effective carbon rates in France consisted primarily of specific taxes on energy use and to a lesser extent of permit prices from the EU ETS. France priced 75% of its energy related CO_2 emissions, and 34% were priced above EUR 30 per tonne of CO_2 (see Figure 6.27); the vast majority of these emissions was from road transport. The coverage of taxes and ETS of emissions from the industry sector overlapped significantly, with 47% of priced emissions in the industry sector subject to both pricing instruments (see Figure 6.28). The majority of unpriced emissions were emitted by industry, as well as by the residential and commercial sector.

France recently introduced a carbon component in existing taxes on energy, the level of which is set to increase over time. As explained in Annex A, the price signal from tax policies introduced after 2012 is not reflected in this report.

Germany

Figure 6.29. **Proportion of CO_2 emissions from energy use subject to different levels of effective carbon rates in Germany in 2012**

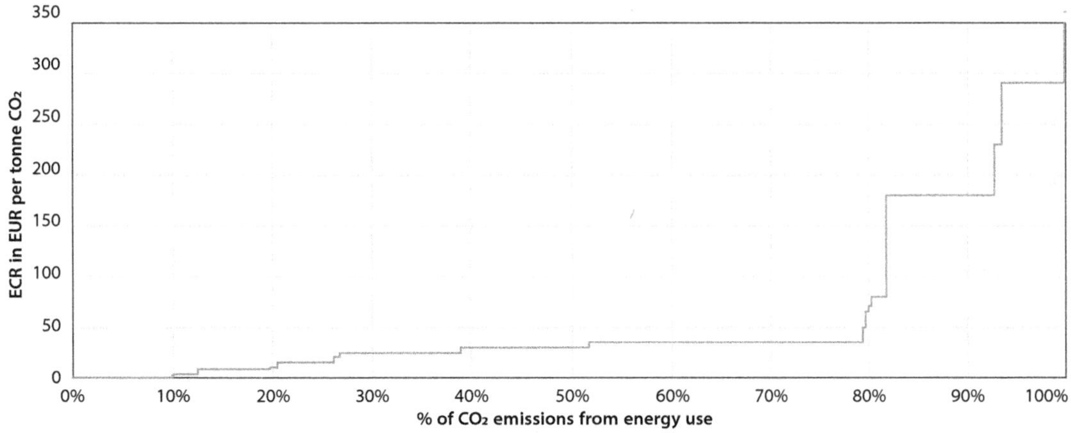

StatLink http://dx.doi.org/10.1787/888933390357

Figure 6.30. **Average effective carbon rates in Germany by sector and component in 2012**

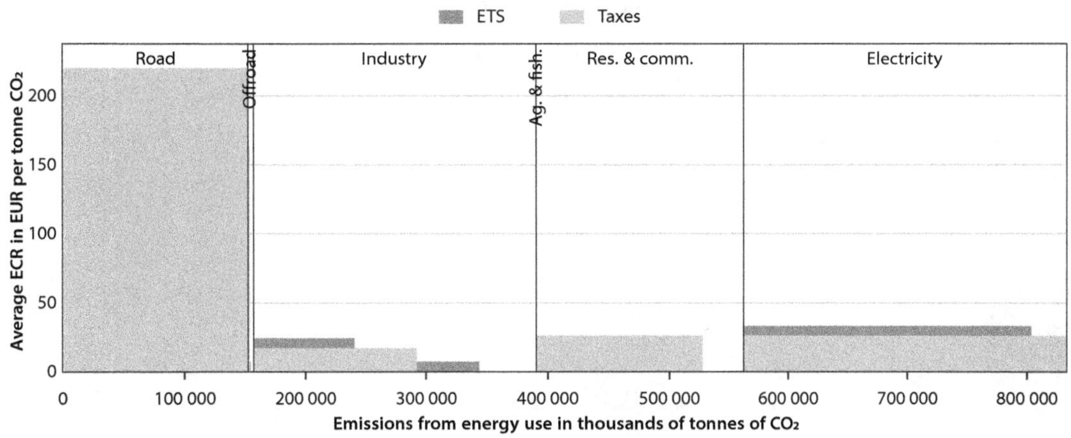

StatLink http://dx.doi.org/10.1787/888933390776

Note: Note: See Box 3.1 in Chapter 3 for notes on the interpretation of effective carbon rates. Please see Annex A and Annex B for detail on methodology and underlying sources.

In 2012, effective carbon rates in Germany consisted primarily of specific taxes on energy use and to a small extent of permit prices from the EU ETS. Germany did not have an explicit carbon tax. Germany priced 90% of carbon emissions from energy use, and 48% were priced above EUR 30 per tonne of CO_2 (see Figure 6.29); a large share of these emissions was from the road sector. The coverage of taxes and ETS of emissions from the industry sector overlapped significantly, and 45% of priced emissions in the industry sector were subject to both pricing instruments (see Figure 6.30). However, the majority of unpriced emissions were from industry, as well as from the residential and commercial sector.

Greece

Figure 6.31. **Proportion of CO_2 emissions from energy use subject to different levels of effective carbon rates in Greece in 2012**

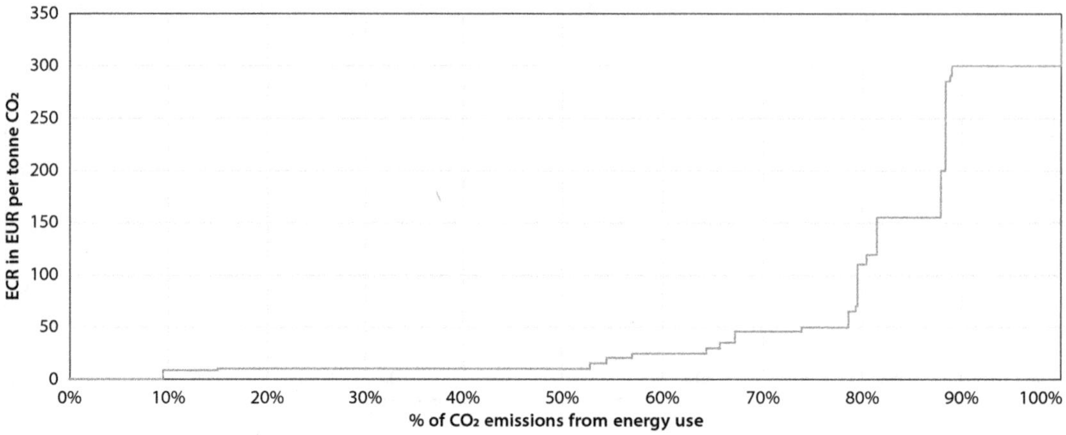

StatLink http://dx.doi.org/10.1787/888933390365

Figure 6.32. **Average effective carbon rates in Greece by sector and component in 2012**

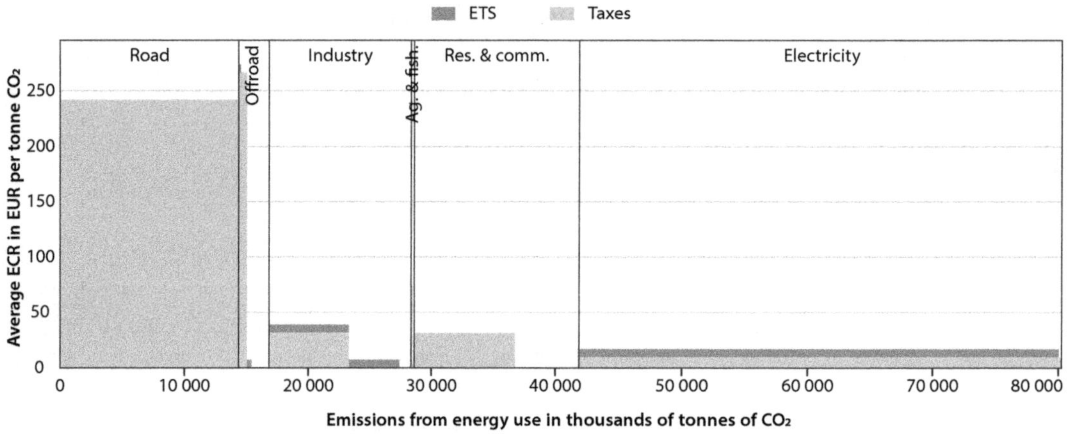

StatLink http://dx.doi.org/10.1787/888933390789

Note: See Box 3.1 in Chapter 3 for notes on the interpretation of effective carbon rates. Please see Annex A and Annex B for detail on methodology and underlying sources.

In 2012, effective carbon rates consisted primarily of specific taxes on energy use and a small part of the ECR consisted of prices from tradable emissions permits. Greece did not have an explicit carbon tax. Greece priced 91% of carbon emissions from energy use, and 34% were priced above EUR 30 per tonne of CO_2 (see Figure 6.31); a large share of these emissions was from the road sector. Unpriced emissions were found primarily in the residential and commercial sector. The price from tradable emissions permits was a significant component of the ECR in the electricity sector (see Figure 6.32).

Hungary

Figure 6.33. **Proportion of CO_2 emissions from energy use subject to different levels of effective carbon rates in Hungary in 2012**

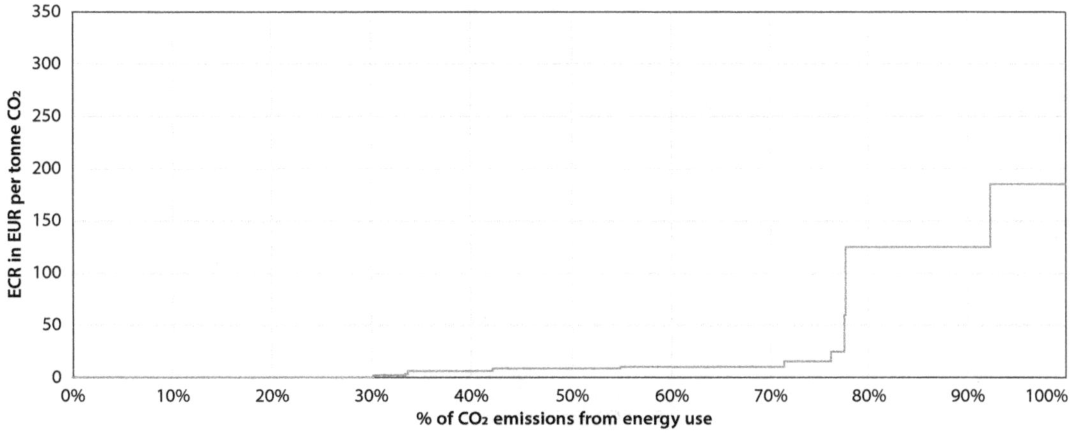

StatLink http://dx.doi.org/10.1787/888933390377

Figure 6.34. **Average effective carbon rates in Hungary by sector and component in 2012**

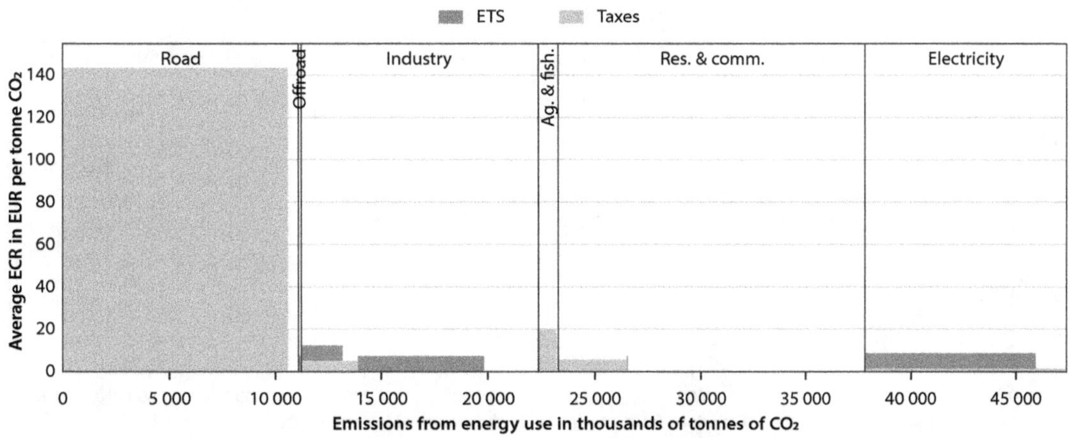

StatLink http://dx.doi.org/10.1787/888933390790

Note: See Box 3.1 in Chapter 3 for notes on the interpretation of effective carbon rates. Please see Annex A and Annex B for detail on methodology and underlying sources.

In 2012, effective carbon rates in Hungary consisted primarily of specific taxes on energy use and to a smaller extent of permit prices from the EU ETS. Hungary did not have an explicit carbon tax. Hungary priced 70% of its carbon emissions from energy use, and 22% were priced above EUR 30 per tonne of CO_2; (Figure 6.33). All of these emissions were from the road sector. Within the industry sector, taxes and the EU ETS applied to a largely separate emissions base. Carbon prices from tradable emissions permits were a significant component of the ECR in the electricity sector. The majority of unpriced emissions were from industry, as well as from the residential and commercial sector (Figure 6.34).

Iceland

Figure 6.35. **Proportion of CO_2 emissions from energy use subject to different levels of effective carbon rates in Iceland in 2012**

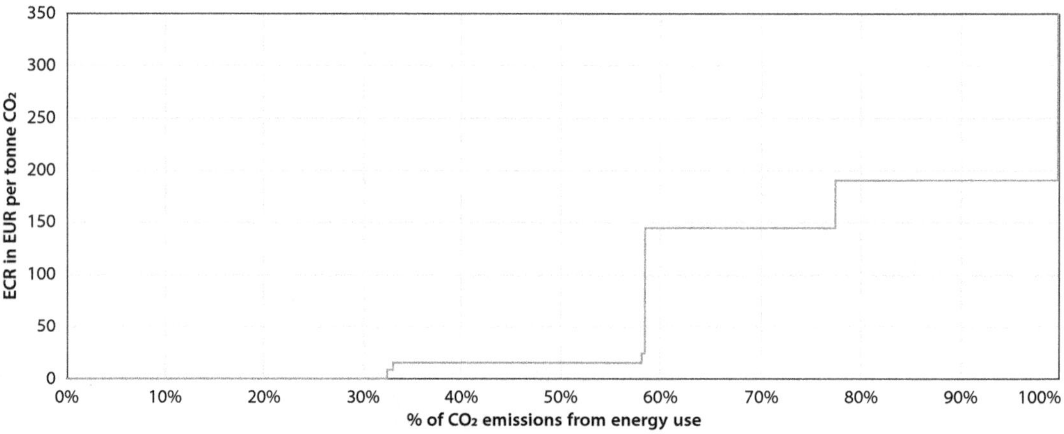

StatLink http://dx.doi.org/10.1787/888933390386

Figure 6.36. **Average effective carbon rates in Iceland by sector and component in 2012**

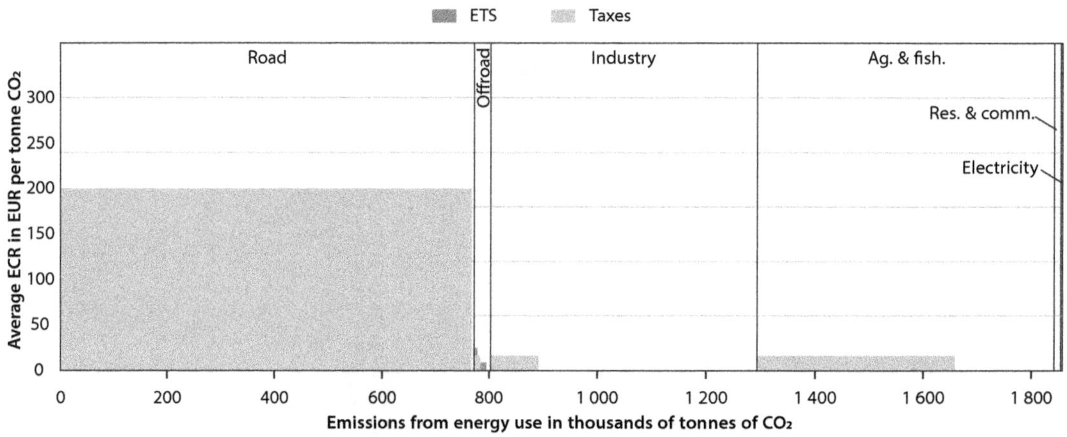

StatLink http://dx.doi.org/10.1787/888933390807

Note: See Box 3.1 in Chapter 3 for notes on the interpretation of effective carbon rates. Please see Annex A and Annex B for detail on methodology and underlying sources.

In 2012, effective carbon rates in Iceland consisted primarily of specific taxes on energy use and to a smaller extent of national carbon taxes and permit prices from the EU ETS. Iceland priced 68% of carbon emissions from energy use, and 42% were priced above EUR 30 per tonne of CO_2 (Figure 6.35); all of these emissions are from the road transport. The majority of unpriced emissions were emitted by the industry sector, as well as by agriculture and fisheries (Figure 6.36).

India

Figure 6.37. **Proportion of CO_2 emissions from energy use subject to different levels of effective carbon rates in India in 2012**

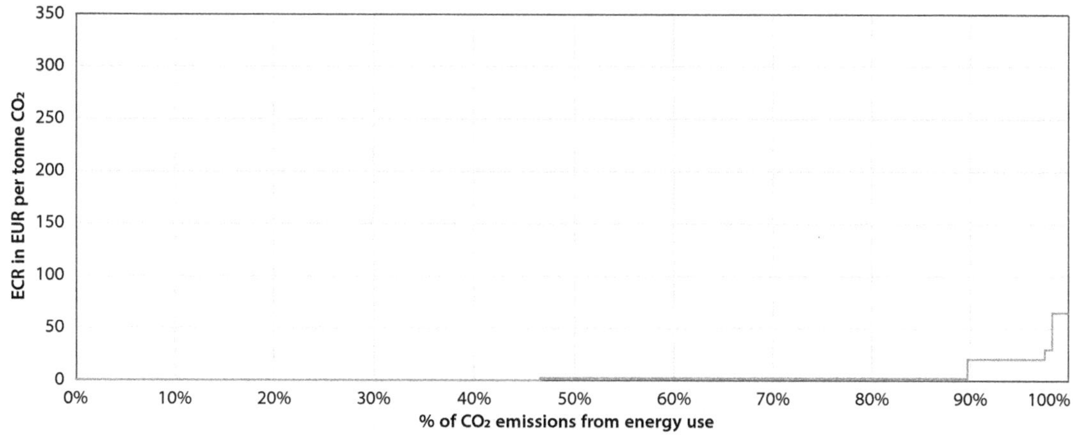

StatLink http://dx.doi.org/10.1787/888933390392

Figure 6.38. **Average effective carbon rates in India by sector and component in 2012**

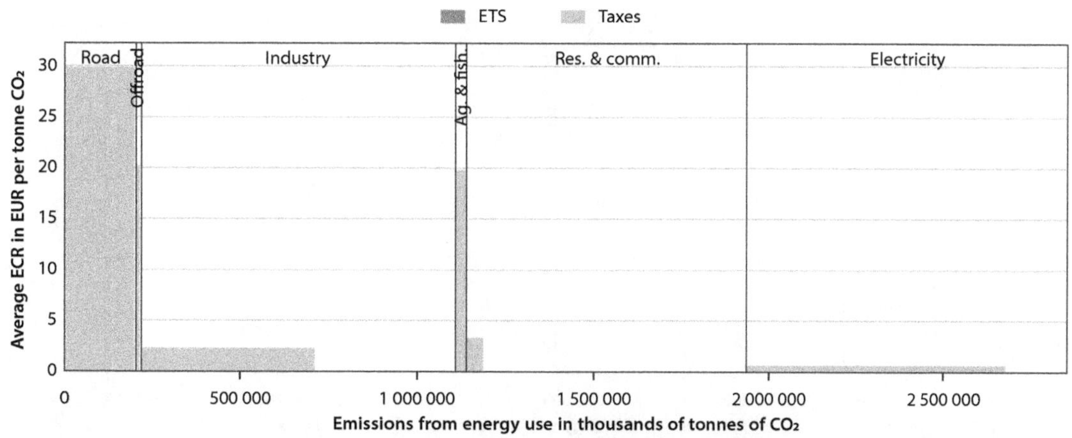

StatLink http://dx.doi.org/10.1787/888933390815

Note: See Box 3.1 in Chapter 3 for notes on the interpretation of effective carbon rates. Please see Annex A and Annex B for detail on methodology and underlying sources.

In 2012, effective carbon rates in India consisted entirely of specific taxes on energy use. India did not have an explicit carbon tax or an emissions trading system for CO_2 emissions. India priced 53% of carbon emissions from energy use, and 2% were priced above EUR 30 per tonne of CO_2 (Figure 6.37); all of these were from road transport. The majority of unpriced emissions were from the industry, residential and commercial sectors (Figure 6.38). Since 2012, India has increased tax rates on fuels for road transport and on coal, which have significantly increased the price signals from these instruments relative to 2012.

Indonesia

Figure 6.39. **Proportion of CO_2 emissions from energy use subject to different levels of effective carbon rates in Indonesia in 2012**

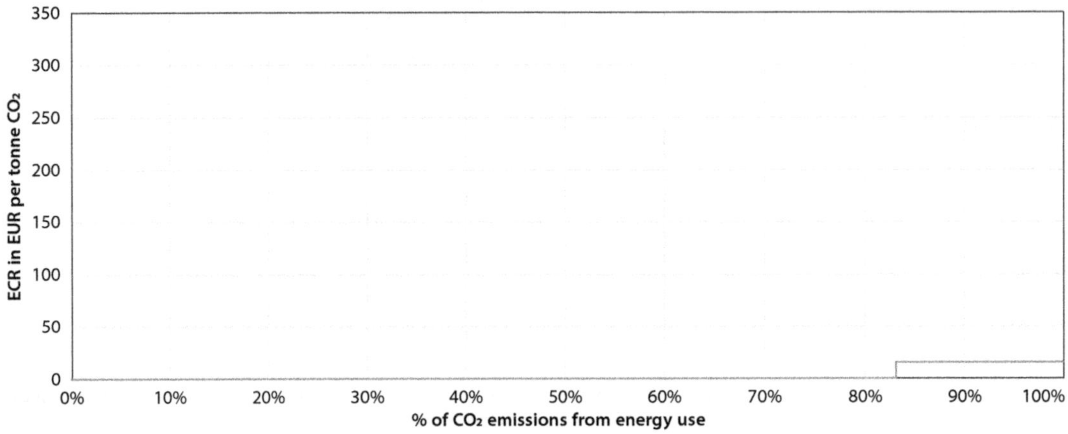

StatLink http://dx.doi.org/10.1787/888933390408

Figure 6.40. **Average effective carbon rates in Indonesia by sector and component in 2012**

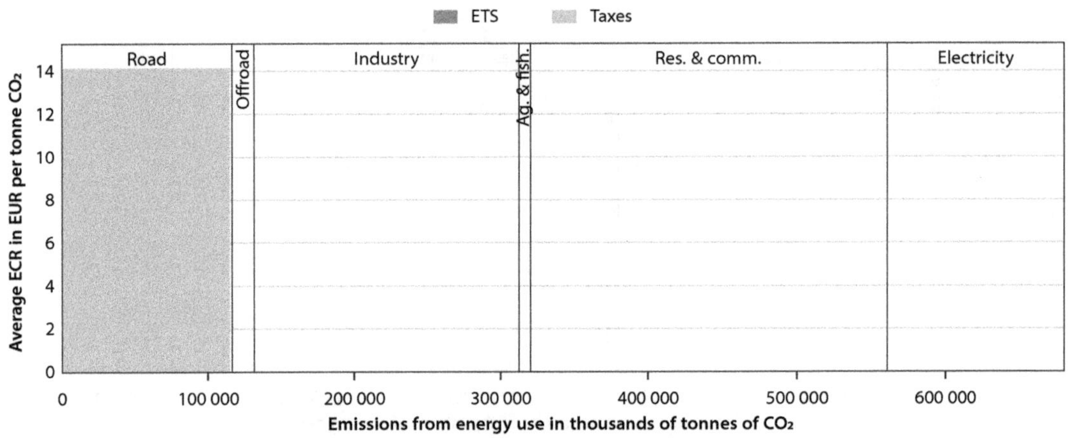

StatLink http://dx.doi.org/10.1787/888933390820

Note: See Box 3.1 in Chapter 3 for notes on the interpretation of effective carbon rates. Please see Annex A and Annex B for detail on methodology and underlying sources.

In 2012, effective carbon rates in Indonesia consisted entirely of specific taxes on energy use, and they only applied to fuels used in road transport (Figure 6.39). Indonesia did neither have an explicit carbon tax nor an emissions trading system. Indonesia priced 17% of carbon emissions of energy use, and none were priced above EUR 30 per tonne of CO_2 (Figure 6.40).

Ireland

Figure 6.41. **Proportion of CO_2 emissions from energy use subject to different levels of effective carbon rates in Ireland in 2012**

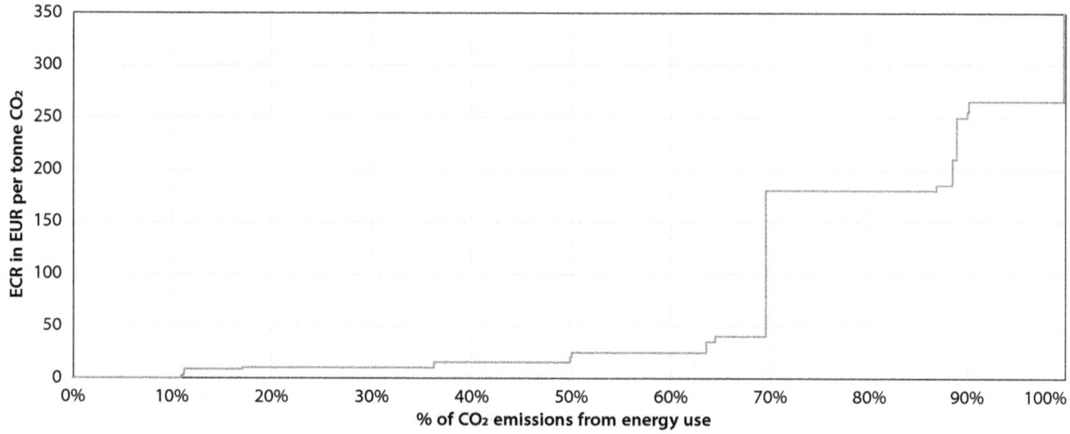

StatLink http://dx.doi.org/10.1787/888933390418

Figure 6.42. **Average effective carbon rates in Ireland by sector and component in 2012**

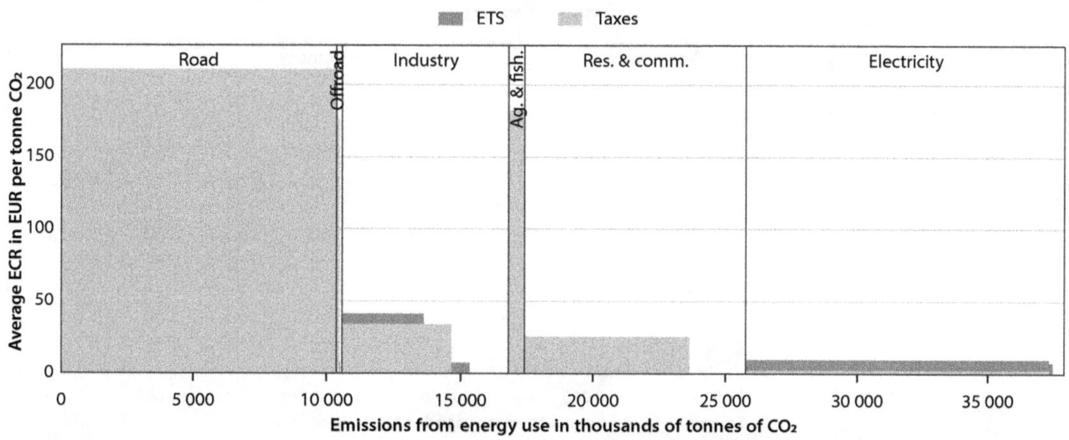

StatLink http://dx.doi.org/10.1787/888933390839

Note: See Box 3.1 in Chapter 3 for notes on the interpretation of effective carbon rates. Please see Annex A and Annex B for detail on methodology and underlying sources.

In 2012, effective carbon rates in Ireland consisted primarily of specific taxes on energy use, and to a smaller extent of national carbon taxes and permit prices from the EU ETS. Ireland priced 89% of carbon emissions from energy use, and 36% were priced above EUR 30 per tonne of CO_2 (Figure 6.41); a large share of these emissions was from road transport. The coverage of taxes and ETS of emissions from the industry sector overlapped significantly, and 65% of priced emissions in the industry sector were subject to both pricing instruments. Prices from tradable emissions permit prices of the EU ETS were a significant component of the ECR in the electricity sector. The majority of unpriced emissions were from industry, as well as from the residential and commercial sector (Figure 6.42).

Israel

Figure 6.43. **Proportion of CO_2 emissions from energy use subject to different levels of effective carbon rates in Israel in 2012**

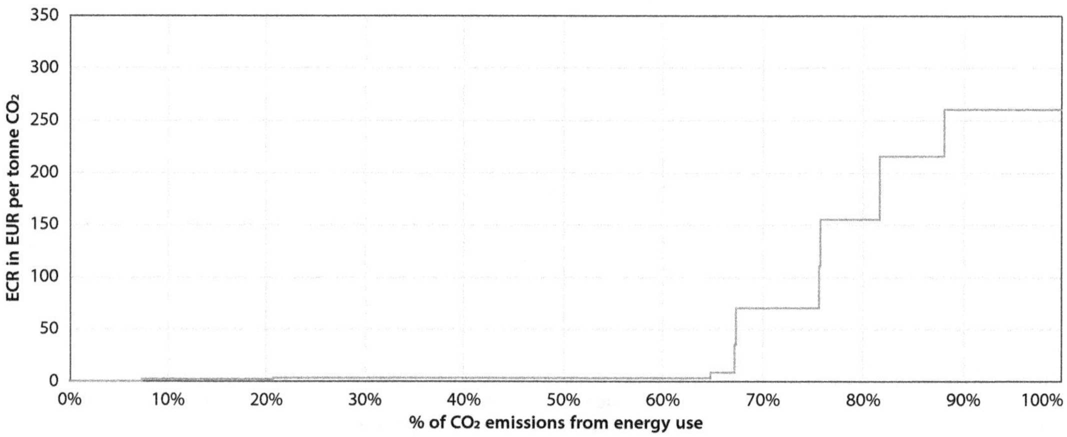

StatLink http://dx.doi.org/10.1787/888933390425

Figure 6.44. **Average effective carbon rates in Israel by sector and component in 2012**

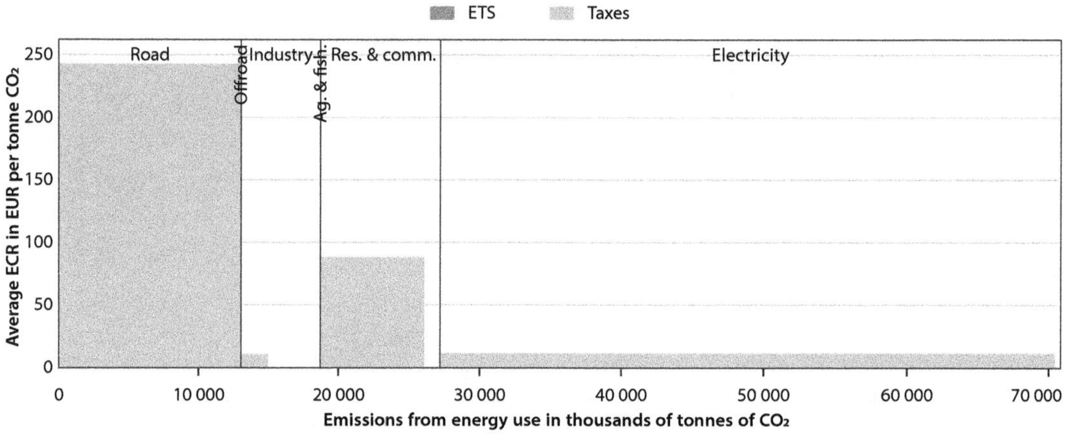

StatLink http://dx.doi.org/10.1787/888933390842

Note: See Box 3.1 in Chapter 3 for notes on the interpretation of effective carbon rates. Please see Annex A and Annex B for detail on methodology and underlying sources.

In 2012, effective carbon rates in Israel consisted entirely of specific taxes on energy use. Israel did not have an explicit carbon tax or an emissions trading system. Israel priced 93% of carbon emissions from energy use, and 33% were priced above EUR 30 per tonne of CO_2 (Figure 6.43); the majority of these emissions were from road transport. The share of unpriced emissions was very small, and the majority of it was from industry (Figure 6.44).

Italy

Figure 6.45. **Proportion of CO_2 emissions from energy use subject to different levels of effective carbon rates in Italy in 2012**

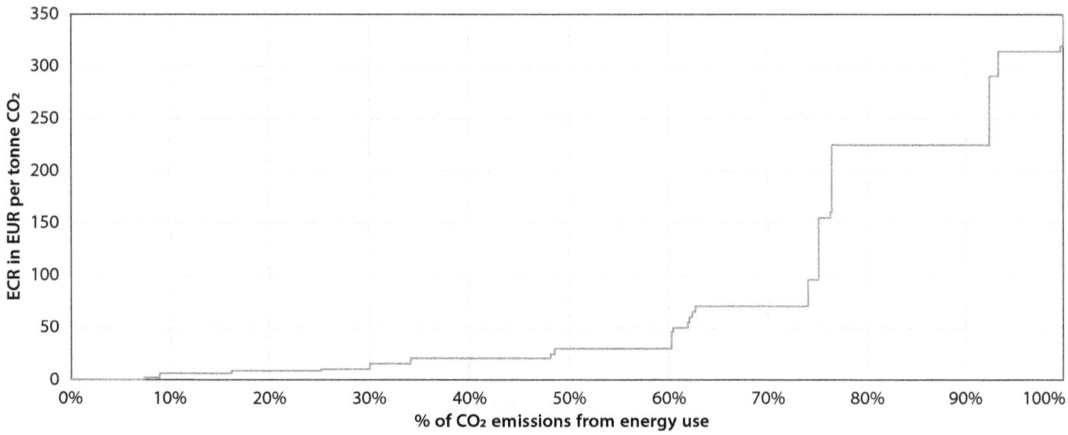

StatLink http://dx.doi.org/10.1787/888933390433

Figure 6.46. **Average effective carbon rates in Italy by sector and component in 2012**

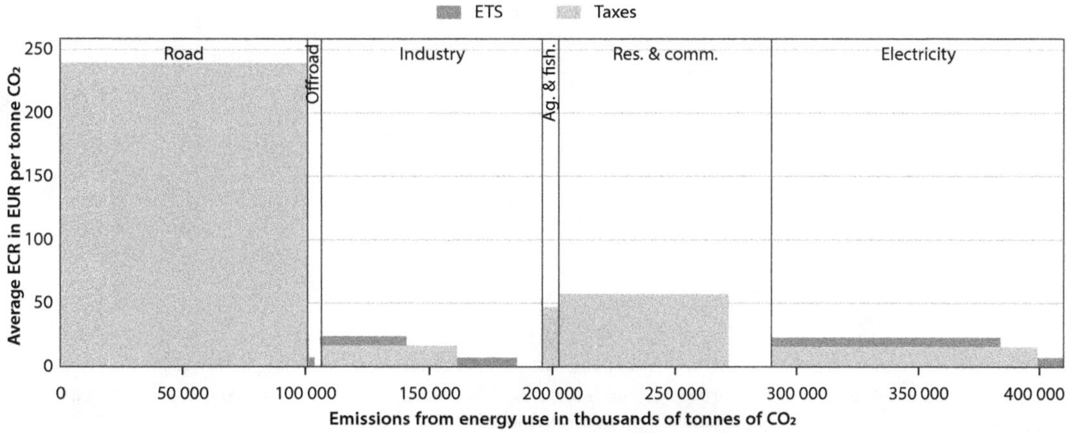

StatLink http://dx.doi.org/10.1787/888933390859

Note: See Box 3.1 in Chapter 3 for notes on the interpretation of effective carbon rates. Please see Annex A and Annex B for detail on methodology and underlying sources.

In 2012, effective carbon rates in Italy consisted primarily of specific taxes on energy use, and to a smaller extent of permit prices from the EU ETS. Italy did not have an explicit carbon tax. Italy priced 93% of carbon emissions from energy use, and 40% were priced above EUR 30 per tonne of CO_2 (Figure 6.45); a large share of these emissions was from road transport. Taxes and prices from tradable emissions permits overlapped substantially in the industry and electricity sector (Figure 6.46).

Japan

Figure 6.47. **Proportion of CO_2 emissions from energy use subject to different levels of effective carbon rates in Japan in 2012**

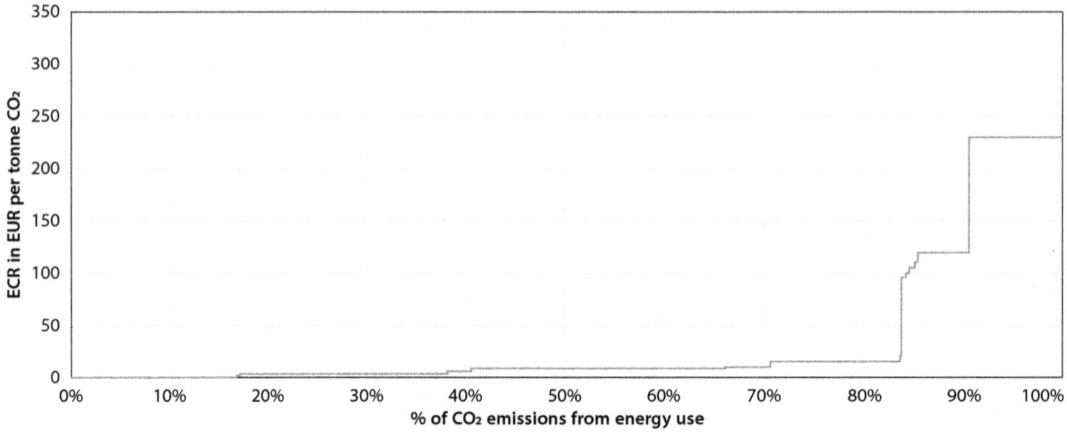

StatLink http://dx.doi.org/10.1787/888933390447

Figure 6.48. **Average effective carbon rates in Japan by sector and component in 2012**

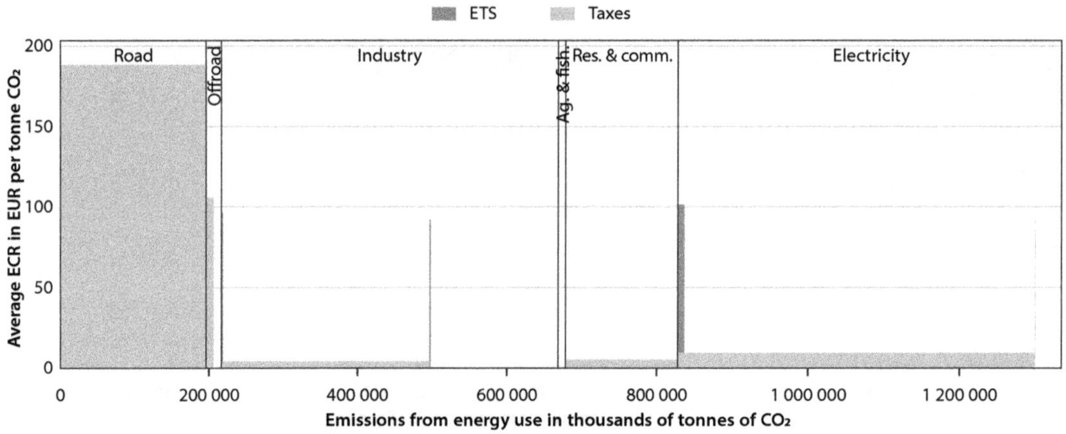

StatLink http://dx.doi.org/10.1787/888933390862

Note: See Box 3.1 in Chapter 3 for notes on the interpretation of effective carbon rates. Please see Annex A and Annex B for detail on methodology and underlying sources.

In 2012, effective carbon rates in Japan consisted primarily of specific taxes on energy use. The Tokyo municipality and the Saitama province operated emissions trading systems, but their emissions base was small when evaluating the emissions of the country as a whole. Japan priced 83% of carbon emissions from energy use, and 16% were priced above EUR 30 per tonne of CO_2 (Figure 6.47). The majority of these emissions were from road transport (Figure 6.48).

Korea

Figure 6.49. **Proportion of CO_2 emissions from energy use subject to different levels of effective carbon rates in Korea in 2012**

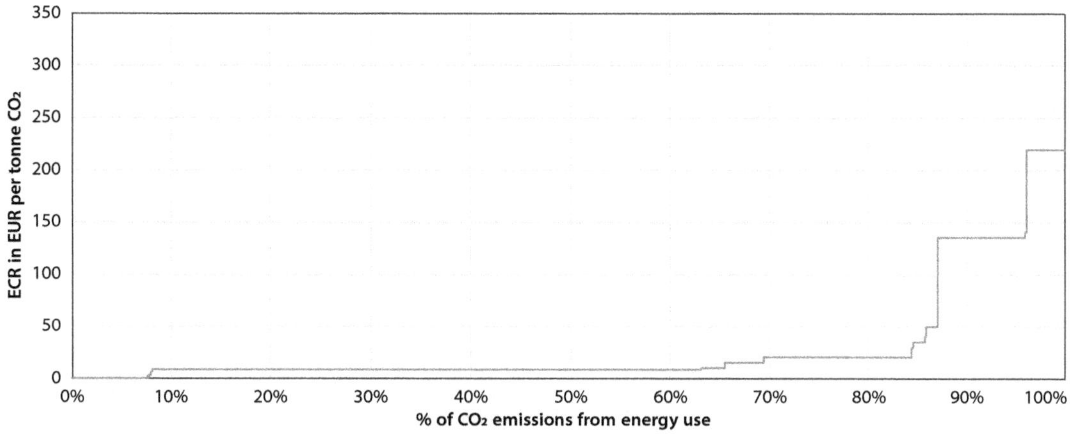

StatLink http://dx.doi.org/10.1787/888933390455

Figure 6.50. **Average effective carbon rates in Korea by sector and component in 2012**

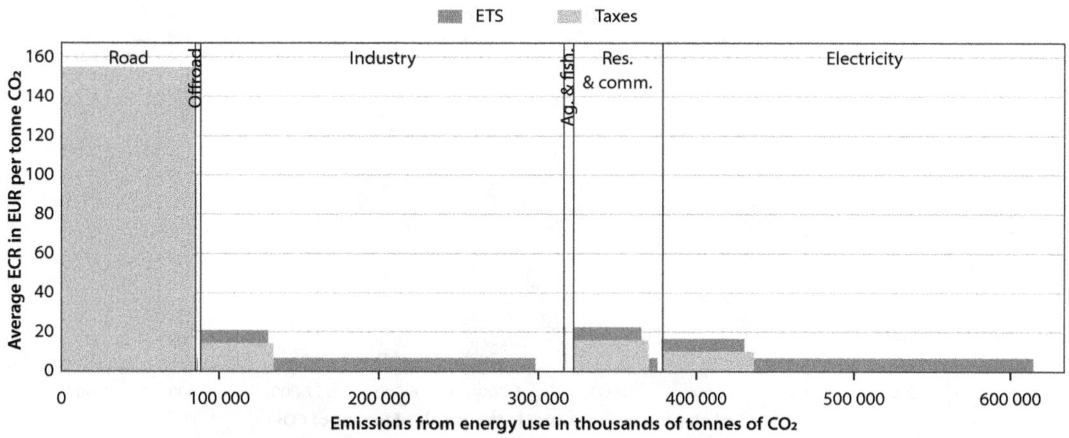

StatLink http://dx.doi.org/10.1787/888933390873

Note: See Box 3.1 in Chapter 3 for notes on the interpretation of effective carbon rates. Please see Annex A and Annex B for detail on methodology and underlying sources.

In 2012, effective carbon rates in Korea consisted of specific taxes on energy use, primarily in the road sector, and permit prices from the ETS, primarily in the industry, residential and commercial, and the electricity sectors.[3] Korea did not have an explicit carbon tax. Korea priced 92% of carbon emissions from energy use, and 16% were priced above EUR 30 per tonne of CO_2 (Figure 6.49). The majority of these emissions were from road transport. The emissions base of the Korea ETS was relatively broad, and it covered around 70% of emissions of the industry and the electricity sector each. Given the broad coverage of the ETS, the overlap between the emissions base of the ETS and taxes was relatively small, at 18% in the industry sector. Permit prices from the Korea ETS were the main component of the ECRs in the industry and electricity sectors (Figure 6.50).

Luxembourg

Figure 6.51. **Proportion of CO_2 emissions from energy use subject to different levels of effective carbon rates in Luxembourg in 2012**

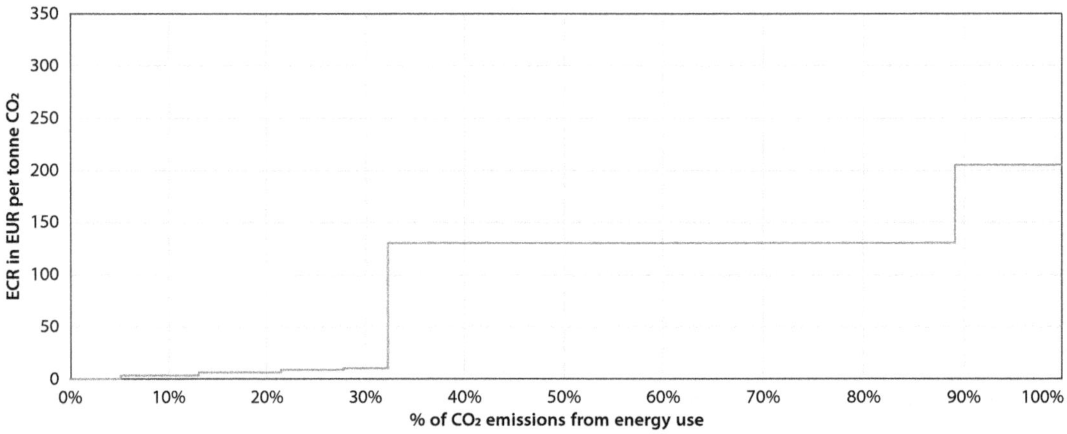

StatLink http://dx.doi.org/10.1787/888933390467

Figure 6.52. **Average effective carbon rates in Luxembourg by sector and component in 2012**

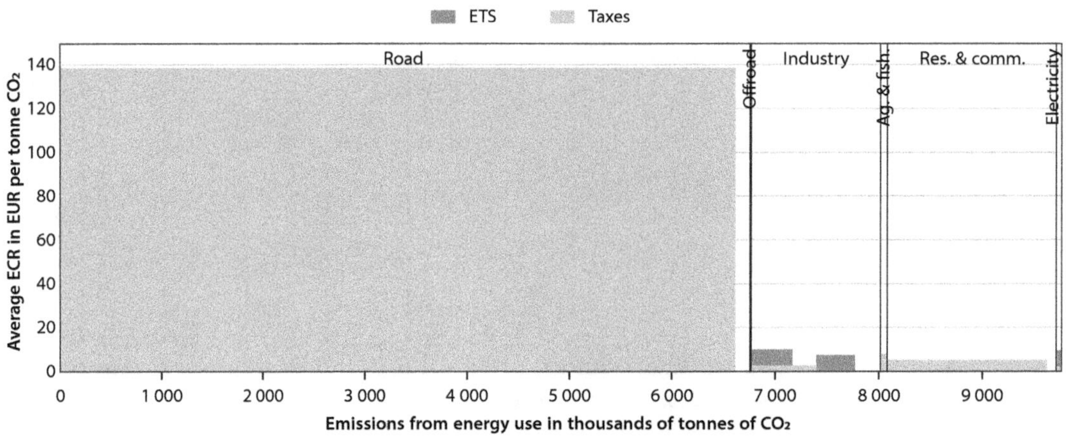

StatLink http://dx.doi.org/10.1787/888933390888

Note: See Box 3.1 in Chapter 3 for notes on the interpretation of effective carbon rates. Please see Annex A and Annex B for detail on methodology and underlying sources.

In 2012, effective carbon rates in Luxembourg were dominated by specific taxes on energy use and permit prices from the EU ETS accounted only for a very small proportion. Luxembourg did not have an explicit carbon tax. Luxembourg priced 95% of carbon emissions from energy use, and 68% were priced above EUR 30 per tonne of CO_2 (Figure 6.51). These emissions were entirely from road transport, which accounted for 70% of all carbon emissions from energy use in Luxembourg. The share of unpriced emissions was relatively small, but stemmed in majority from the industry, and the residential and commercial sector (Figure 6.52). The share of transport fuels in the base was very large, and this was largely because of fuel tourism and fuel purchased by transit traffic, not because of domestic consumption.

Mexico

Figure 6.53. **Proportion of CO_2 emissions from energy use subject to different levels of effective carbon rates in Mexico in 2012**

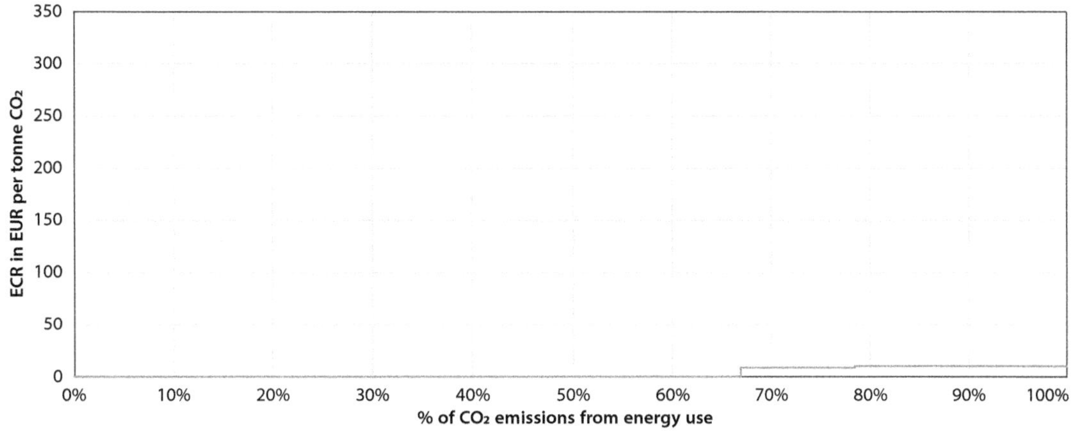

StatLink http://dx.doi.org/10.1787/888933390477

Figure 6.54. **Average effective carbon rates in Mexico by sector and component in 2012**

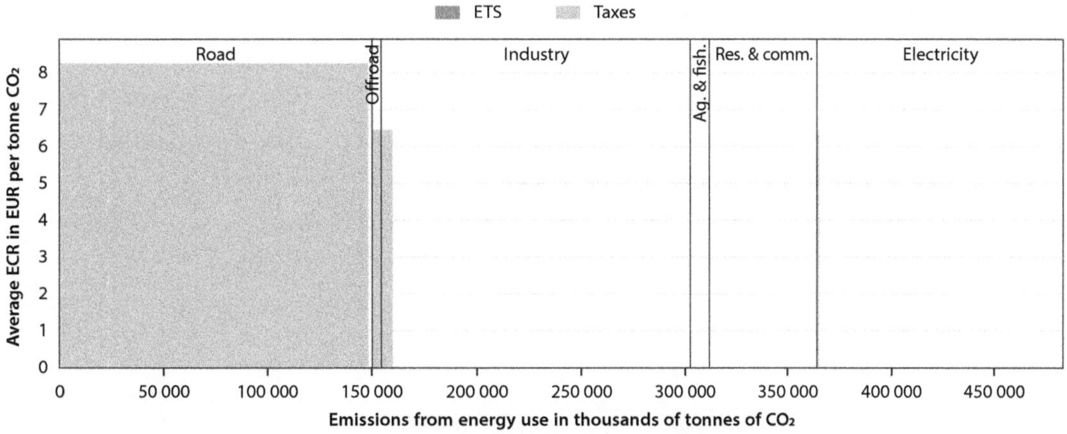

StatLink http://dx.doi.org/10.1787/888933390895

Note: See Box 3.1 in Chapter 3 for notes on the interpretation of effective carbon rates. Please see Annex A and Annex B for detail on methodology and underlying sources.

In 2012, effective carbon rates in Mexico consisted entirely of specific taxes on energy use. Mexico did not have an emissions trading system. Mexico priced 33% of carbon emissions of energy use, and none were priced above EUR 30 per tonne of CO_2 (Figure 6.53). CO_2 emissions from energy use in the agriculture and fisheries sector were not priced. The majority of unpriced emissions were from the industry, the residential and commercial, and the electricity sectors (Figure 6.54).

The effective carbon rates in this publication refer to tax rates of April 2012. Since then, Mexico has implemented an environmental tax reform, which included a reform of the excise tax on road transport fuels and the introduction of a carbon tax. These reforms increase both ECR levels and coverage. More specifically, the reform of the excise tax on road transport fuels substantially increases the effective carbon rate in the transport sector. The newly introduced carbon tax substantially expands the proportion of emissions covered by a price in the industry, agriculture and fishing, residential and commercial, as well as the electricity sector.

These policy changes will be reflected in future updates of the data underlying the effective carbon rates. A more detailed description of the Mexican environmental tax reform will be included in a forthcoming OECD paper.

Netherlands

Figure 6.55. **Proportion of CO_2 emissions from energy use subject to different levels of effective carbon rates in the Netherlands in 2012**

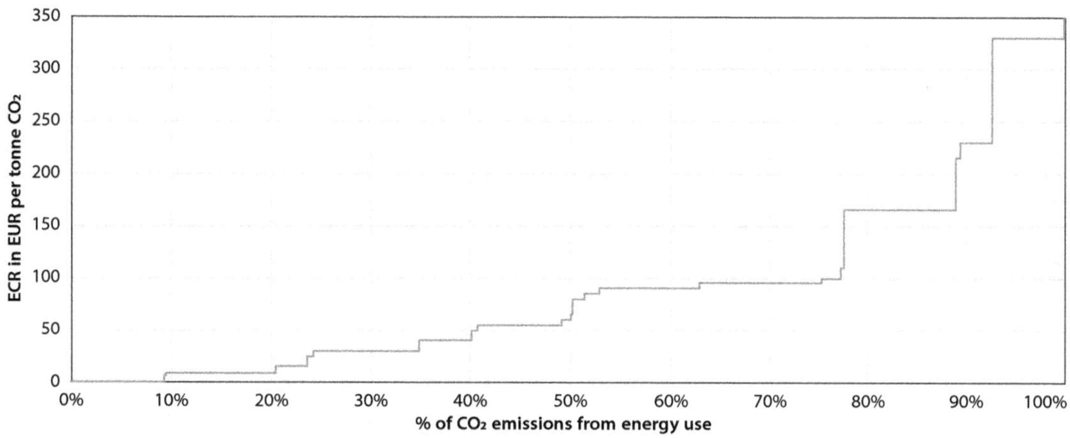

StatLink http://dx.doi.org/10.1787/888933390483

Figure 6.56. **Average effective carbon rates in the Netherlands by sector and component in 2012**

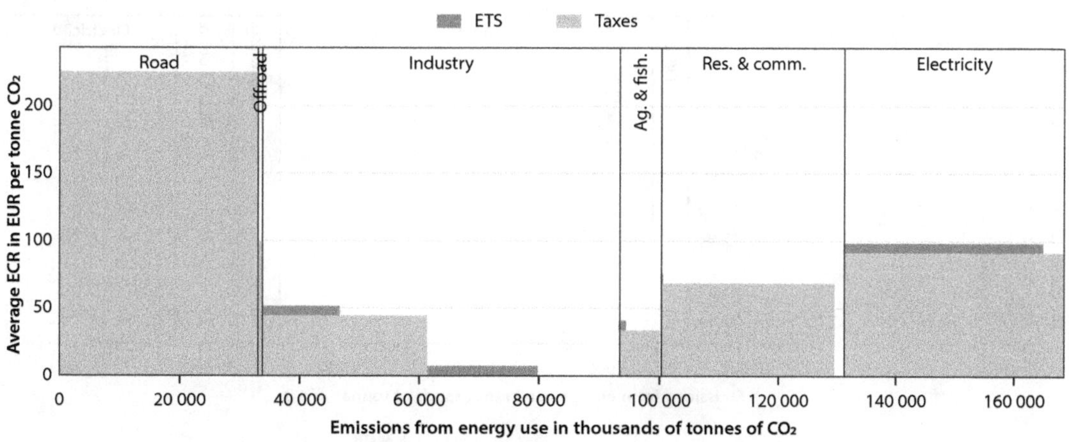

StatLink http://dx.doi.org/10.1787/888933390904

Note: See Box 3.1 in Chapter 3 for notes on the interpretation of effective carbon rates. Please see Annex A and Annex B for detail on methodology and underlying sources.

In 2012, effective carbon rates in the Netherlands consisted primarily of specific taxes on energy use and to a very small extent of permit prices from the EU ETS. The Netherlands did not have an explicit carbon tax. The Netherlands priced 91% of carbon emissions from energy use, and 65% were priced above EUR 30 per tonne of CO_2 (Figure 6.55). While the coverage of carbon prices from tradable emissions permits and taxes overlapped entirely in the electricity sector, the overlap between the two instruments was at 22% of the emissions base in the industry sector. A small share of emissions was unpriced in the Netherlands, the majority of which were from the industry sector (Figure 6.56).

New Zealand

Figure 6.57. **Proportion of CO_2 emissions from energy use subject to different levels of effective carbon rates in New Zealand in 2012**

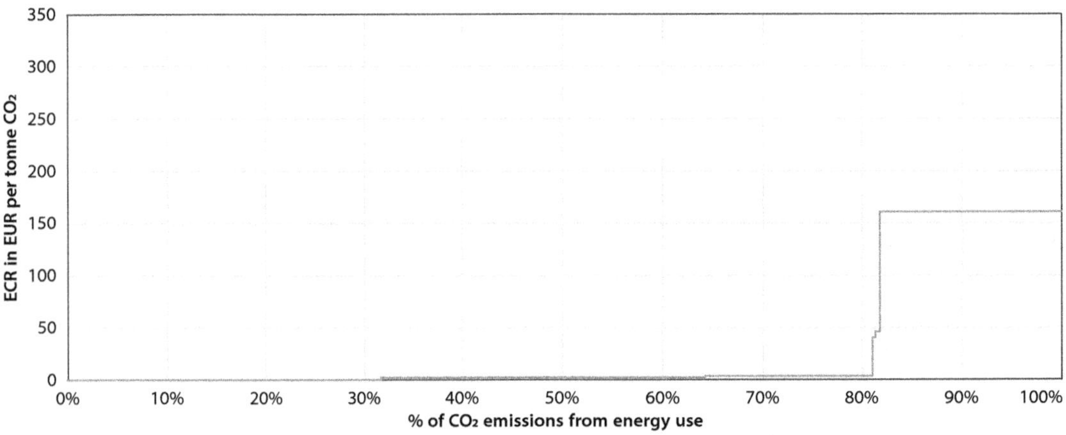

StatLink http://dx.doi.org/10.1787/888933390490

Figure 6.58. **Average effective carbon rates in New Zealand by sector and component in 2012**

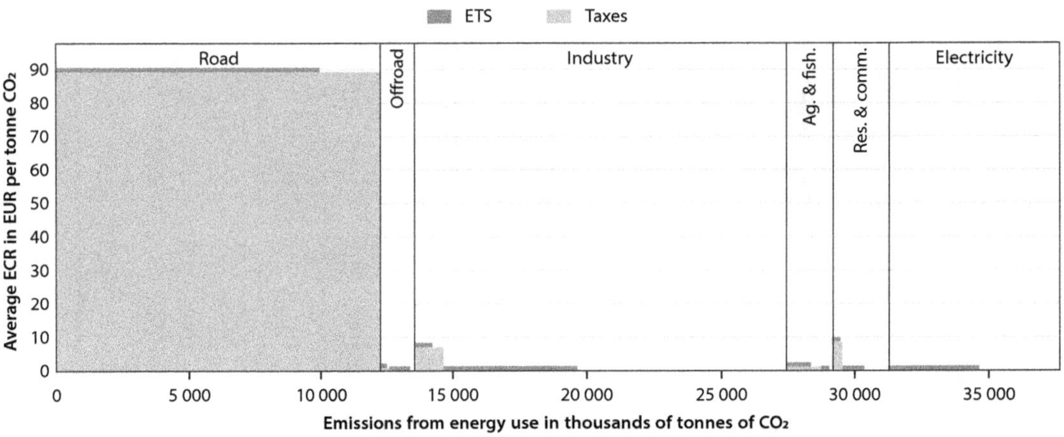

StatLink http://dx.doi.org/10.1787/888933390910

Note: See Box 3.1 in Chapter 3 for notes on the interpretation of effective carbon rates. Please see Annex A and Annex B for detail on methodology and underlying sources.

In 2012, effective carbon in New Zealand rates consisted primarily of specific taxes on energy use and to a small extent of permit prices from the New Zealand ETS. New Zealand did not have an explicit carbon tax. New Zealand priced 68% of its carbon emissions from energy use, and 19% were priced above EUR 30 per tonne of CO_2 (Figure 6.57); the vast majority of these emissions were from the road sector. The coverage of taxes and permit prices in New Zealand was largely separate in the industry and electricity sectors. The ECR on emissions from the electricity sector consisted entirely of carbon prices from tradable permits. Unpriced emissions were primarily from the industry and the electricity sectors (Figure 6.58).

Norway

Figure 6.59. **Proportion of CO_2 emissions from energy use subject to different levels of effective carbon rates in Norway in 2012**

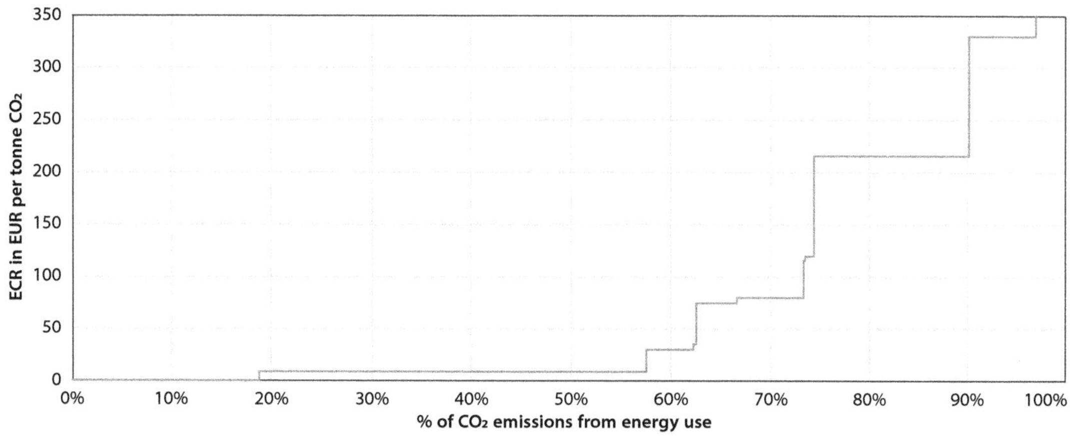

StatLink http://dx.doi.org/10.1787/888933390506

Figure 6.60. **Average effective carbon rates in Norway by sector and component in 2012**

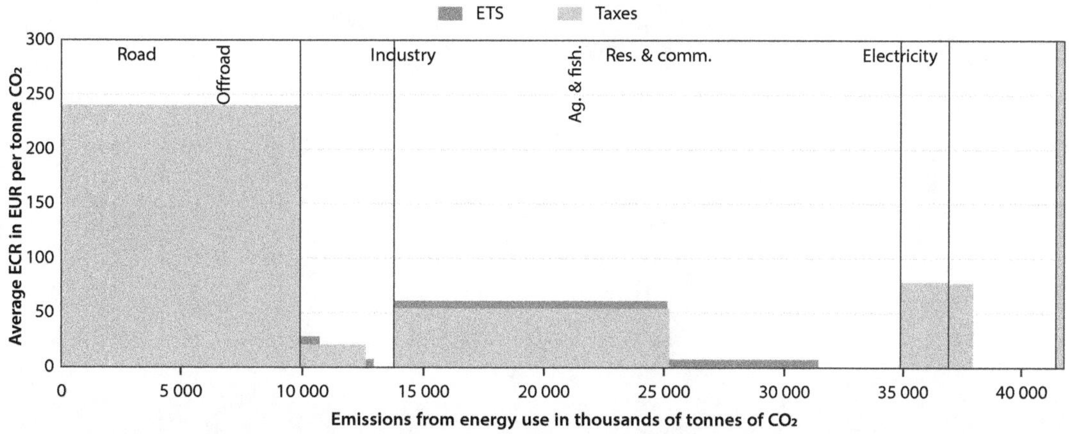

StatLink http://dx.doi.org/10.1787/888933390924

Note: See Box 3.1 in Chapter 3 for notes on the interpretation of effective carbon rates. Please see Annex A and Annex B for detail on methodology and underlying sources.

In 2012, effective carbon rates in Norway consisted primarily of specific taxes on energy use and to a smaller extent of national carbon taxes and permit prices from the EU ETS. Norway priced 81% of carbon emissions from energy use, and 38% were priced above EUR 30 per tonne of CO_2 (Figure 6.59); a significant share of these emissions was from road transport. While the overlap between taxes and carbon prices from tradable emissions permits was significant in the industry sector (54%), prices from tradable emissions permits covered a substantial part of these emissions as the single pricing instrument (30%). Unpriced emissions were primarily from industry, and the residential and commercial sector (Figure 6.60).

A substantial share of unpriced emissions was from the combustion of biomass. Excluding emissions from biomass, 97% of CO_2 emissions from energy use in Norway were priced at a positive rate and 44% were priced above EUR 30 per tonne of CO_2. In the residential and commercial sector, 74% of CO_2 emissions were priced above EUR 30 per tonne of CO_2 when emissions from biomass were excluded. In the industry sector, all emissions were priced when excluding emissions from biomass, and 14% were priced above EUR 30 per tonne of CO_2.

Poland

Figure 6.61. **Proportion of CO_2 emissions from energy use subject to different levels of effective carbon rates in Poland in 2012**

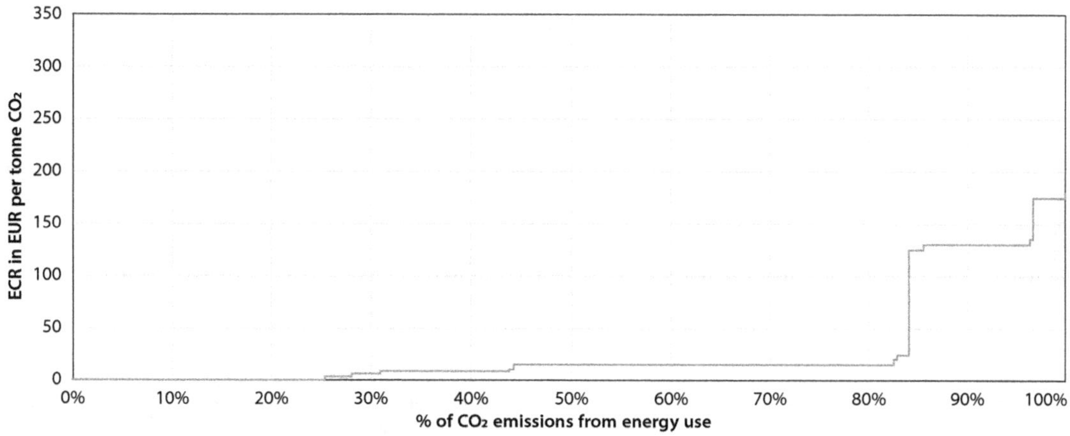

StatLink http://dx.doi.org/10.1787/888933390519

Figure 6.62. **Average effective carbon rates in Poland by sector and component in 2012**

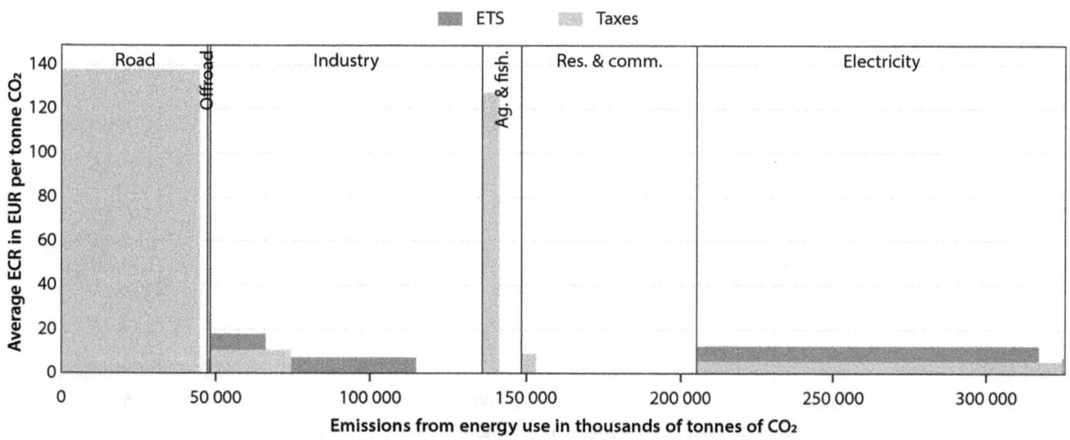

StatLink http://dx.doi.org/10.1787/888933390930

Note: See Box 3.1 in Chapter 3 for notes on the interpretation of effective carbon rates. Please see Annex A and Annex B for detail on methodology and underlying sources.

In 2012, effective carbon rates in Poland consisted primarily of specific taxes on energy use, and to a small extent of permit prices from the EU ETS. Poland did not have an explicit carbon tax. Poland priced 75% of carbon emissions from energy use, and 16% were priced above EUR 30 per tonne of CO_2 (Figure 6.61); a large share of these emissions was from road transport. The coverage of carbon prices from tradable permits and taxes was largely separate in industry. Permit prices from the EU ETS were a significant component of the ECRs in the electricity sector. The majority of unpriced emissions was from industry and the residential and the commercial sector (Figure 6.62).

The effective carbon rates in this publication refer to tax rates in April 2012. Since then, Poland has substantially increased its effective excise tax rates on natural gas, which translates into an increase in the effective carbon rate in different sectors. These policy changes will be reflected in future updates of the energy tax data underlying the effective carbon rates.

Portugal

Figure 6.63. **Proportion of CO_2 emissions from energy use subject to different levels of effective carbon rates in Portugal in 2012**

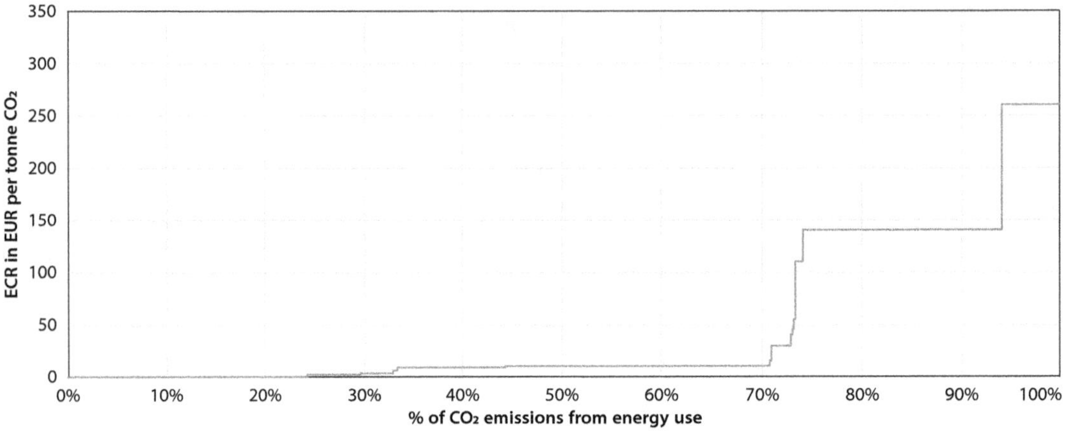

StatLink http://dx.doi.org/10.1787/888933390522

Figure 6.64. **Average effective carbon rates in Portugal by sector and component in 2012**

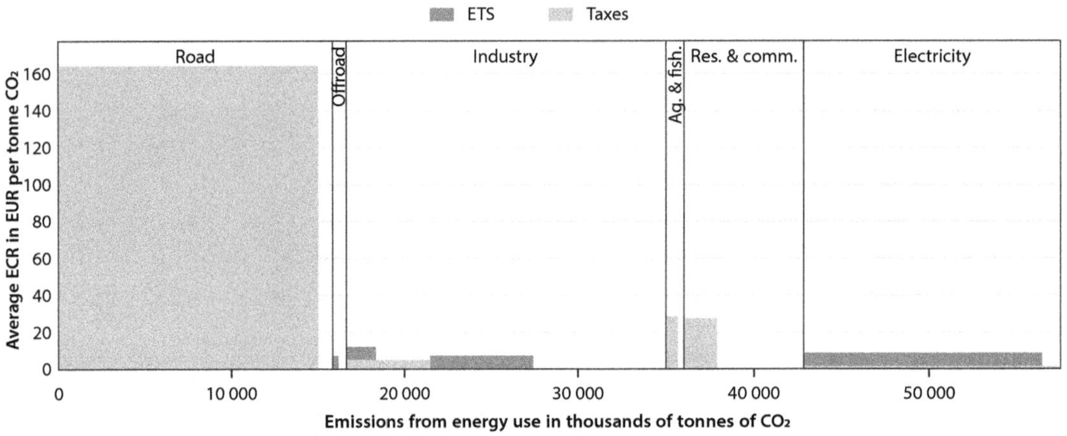

StatLink http://dx.doi.org/10.1787/888933390947

Note: See Box 3.1 in Chapter 3 for notes on the interpretation of effective carbon rates. Please see Annex A and Annex B for detail on methodology and underlying sources.

In 2012, effective carbon rates in Portugal consisted primarily of specific taxes on energy use and to a small extent of permit prices from the EU ETS. Portugal did not have an explicit carbon tax. Portugal priced 76% of carbon emissions from energy use, and 27% were priced above EUR 30 per tonne of CO_2 (Figure 6.63); the vast majority of these emissions were from the road sector. The coverage of taxes and prices from tradable permits was largely separate in the industry sector. The EU ETS applied to 32% of emissions in that sector, and taxes covered 17%. Taxes and ETS applied jointly to 9% of emissions in the industrial sector. Carbon prices from tradable permits were also a significant part of the ECR in the electricity sector. The majority of unpriced emissions were from industry and the residential and commercial sector (Figure 6.64).

In 2015, Portugal introduced a carbon tax. This is not included in the calculation of effective carbon rates as they include taxes in force as of April 2012.

Russian Federation

Figure 6.65. **Proportion of CO_2 emissions from energy use subject to different levels of effective carbon rates in Russia in 2012**

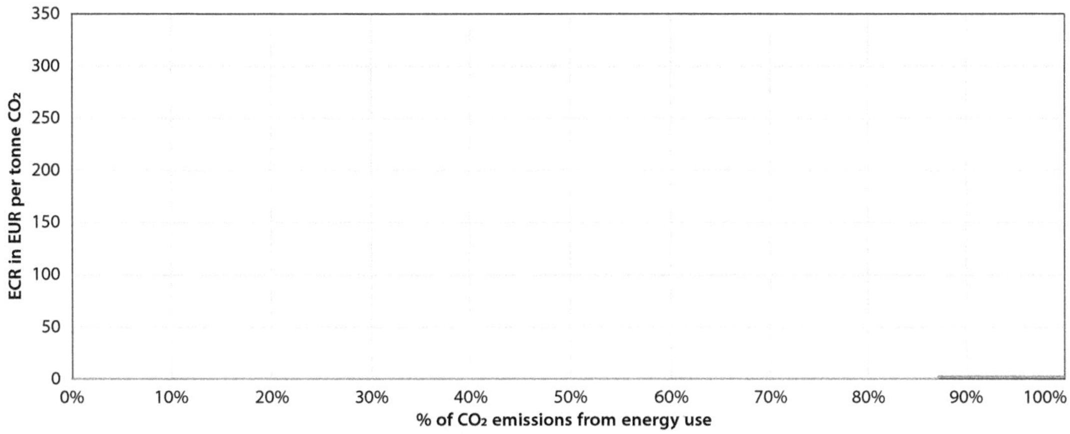

StatLink http://dx.doi.org/10.1787/888933390536

Figure 6.66. **Average effective carbon rates in Russia by sector and component in 2012**

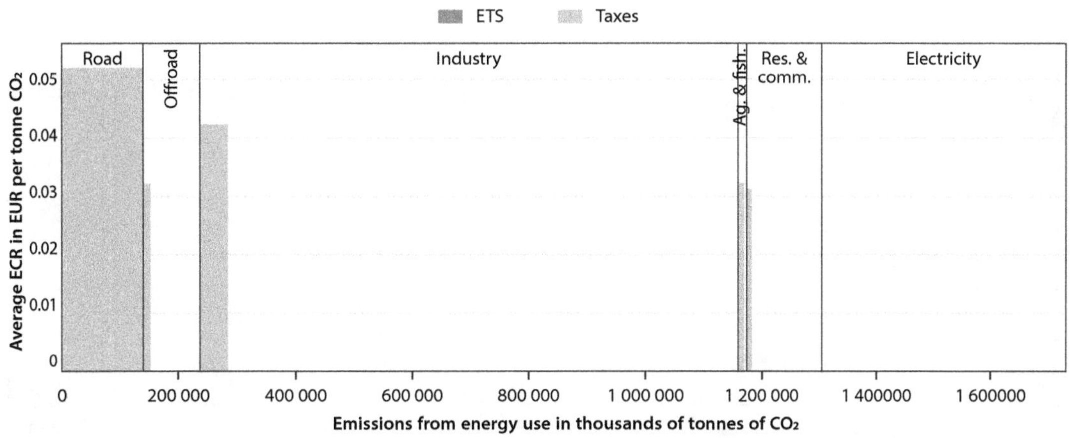

StatLink http://dx.doi.org/10.1787/888933390955

Note: See Box 3.1 in Chapter 3 for notes on the interpretation of effective carbon rates. Please see Annex A and Annex B for detail on methodology and underlying sources.

In 2012, effective carbon rates in Russia consisted entirely of specific taxes on energy use. Russia did not have an explicit carbon tax or an emissions trading system. Russia priced 13% of its energy related CO_2 emissions, none of which were priced above EUR 30 per tonne of CO_2 (Figure 6.65). The majority of emissions from energy use in Russia were unpriced. Most unpriced emissions were from industry, the residential and commercial, and the electricity sectors (Figure 6.66).

Slovak Republic

Figure 6.67. **Proportion of CO_2 emissions from energy use subject to different levels of effective carbon rates in the Slovak Republic in 2012**

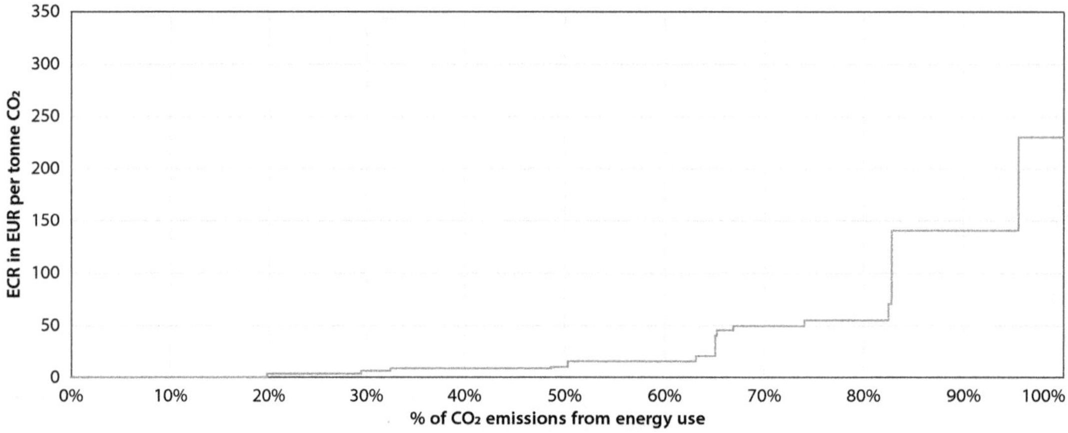

StatLink http://dx.doi.org/10.1787/888933390540

Figure 6.68. **Average effective carbon rates in the Slovak Republic by sector and component in 2012**

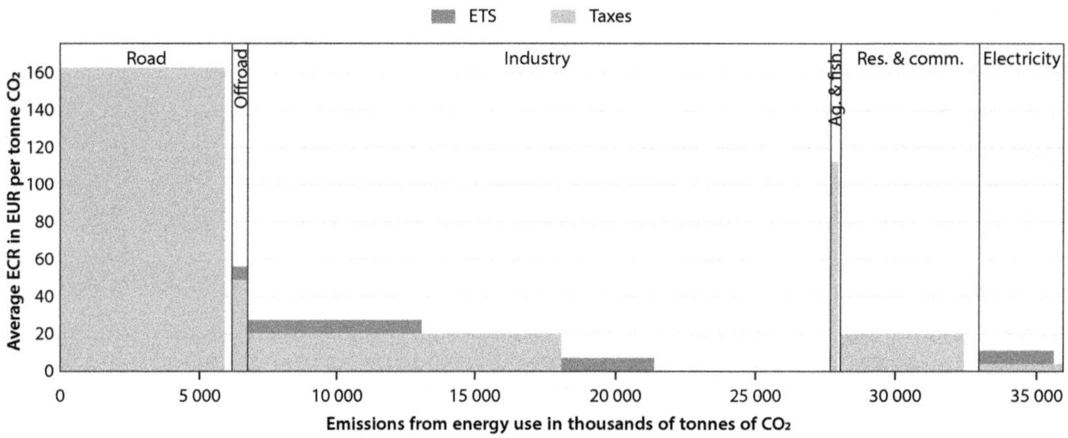

StatLink http://dx.doi.org/10.1787/888933390966

Note: See Box 3.1 in Chapter 3 for notes on the interpretation of effective carbon rates. Please see Annex A and Annex B for detail on methodology and underlying sources.

In 2012, effective carbon rates in the Slovak Republic consisted primarily of specific taxes on energy use and to a small extent of permit prices from the EU ETS. The Slovak Republic did not have an explicit carbon tax. The Slovak Republic priced 80% of carbon emissions from energy use, and 35% were priced above EUR 30 per tonne of CO_2 (Figure 6.67); the majority of these emissions were from road transport. Taxes and carbon prices from tradable permits covered a large part of the emissions base in the industry sector separately. Prices from tradable permits also played a significant part of the effective carbon rates on emissions in the electricity sector. Unpriced emissions were primarily from the industry sector (Figure 6.68).

Slovenia

Figure 6.69. **Proportion of CO_2 emissions from energy use subject to different levels of effective carbon rates in Slovenia in 2012**

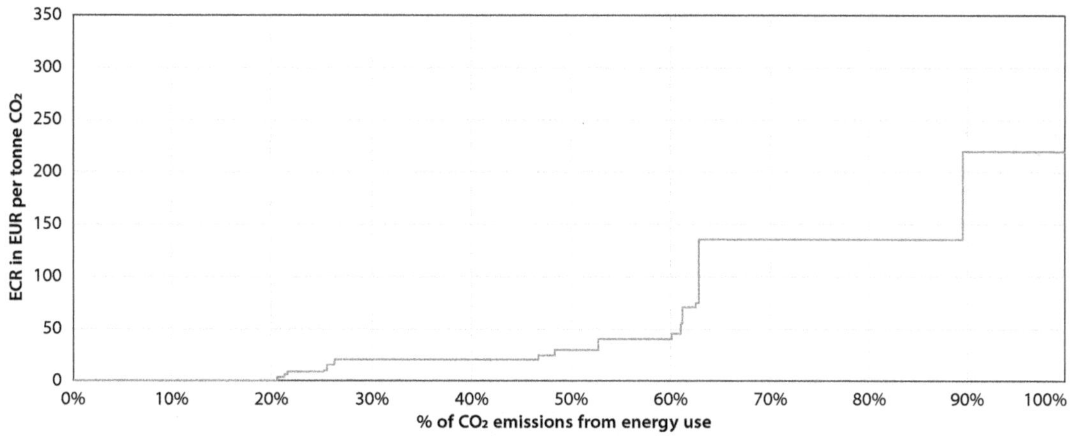

StatLink http://dx.doi.org/10.1787/888933390558

Figure 6.70. **Average effective carbon rates in Slovenia by sector and component in 2012**

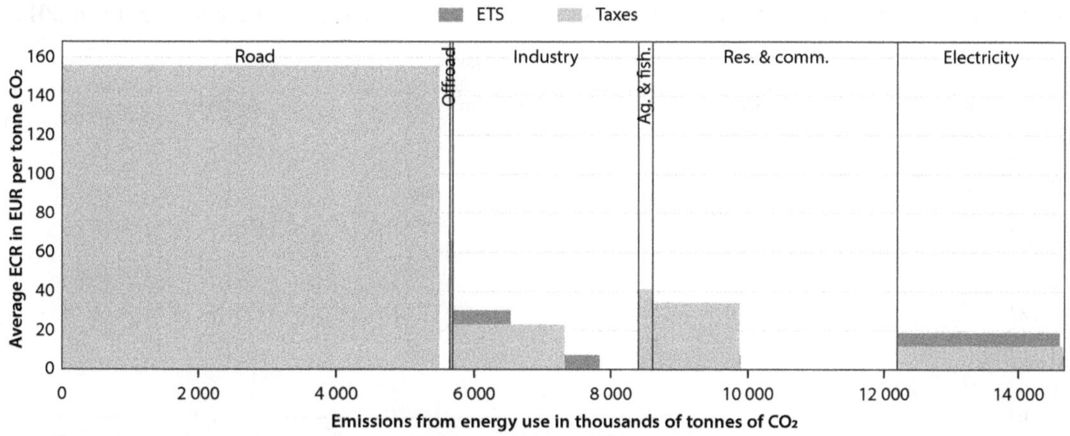

StatLink http://dx.doi.org/10.1787/888933390973

Note: See Box 3.1 in Chapter 3 for notes on the interpretation of effective carbon rates. Please see Annex A and Annex B for detail on methodology and underlying sources.

In 2012, effective carbon rates in Slovenia consisted primarily of specific taxes on energy use and to a lesser extent of national carbon taxes and permit prices from the EU ETS. Slovenia priced 79% of carbon emissions from energy use, and 47% were priced above EUR 30 per tonne of CO_2 (Figure 6.69); the majority of these emissions were from the road sector. Unpriced emissions were primarily from the residential and commercial sector (Figure 6.70).

A substantial share of unpriced emissions was from the combustion of biomass. When excluding emissions from biomass, 97% of CO_2 emissions from energy use in Slovenia were priced at a positive rate, and 58% were priced above EUR 30 per tonne of CO_2. In the residential and commercial sector, 71% of total CO_2 emissions were covered at an ECR above EUR 30 per tonne of CO_2 when excluding emissions from biomass. In the industry sector, 91% of emissions were priced when excluding emissions from biomass and 10% were priced above EUR 30 per tonne of CO_2.

South Africa

Figure 6.71. **Proportion of CO_2 emissions from energy use subject to different levels of effective carbon rates in South Africa in 2012**

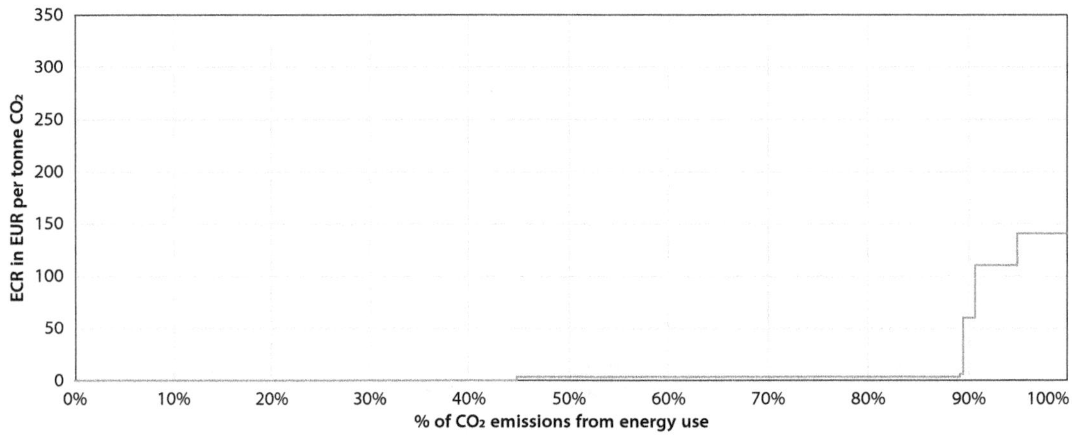

StatLink http://dx.doi.org/10.1787/888933390563

Figure 6.72. **Average effective carbon rates in South Africa by sector and component in 2012**

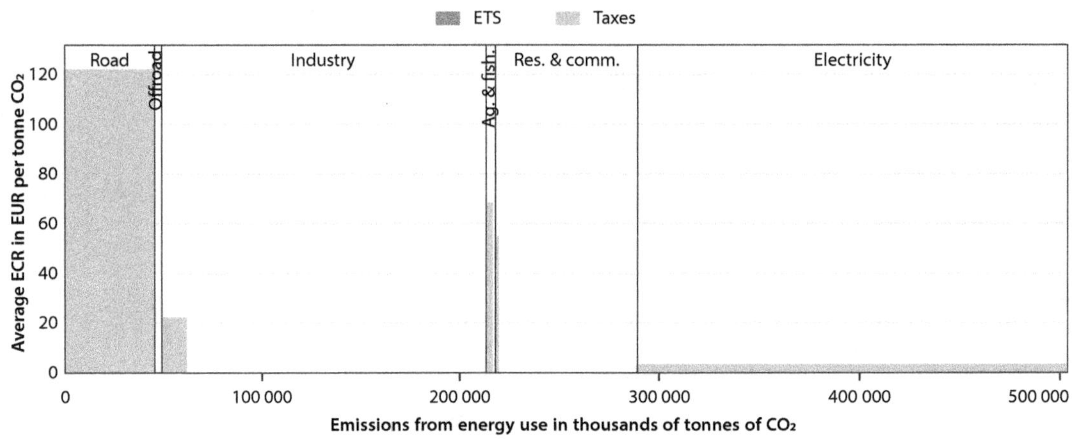

StatLink http://dx.doi.org/10.1787/888933390986

Note: See Box 3.1 in Chapter 3 for notes on the interpretation of effective carbon rates. Please see Annex A and Annex B for detail on methodology and underlying sources.

In 2012, effective carbon rates in South Africa consisted entirely of specific taxes on energy use. South Africa did not have an explicit carbon tax or an emissions trading system. South Africa priced 55% of carbon emissions from energy use, and 11% were priced above EUR 30 per tonne of CO_2 (Figure 6.71). The great majority of these emissions were from the road sector. Unpriced emissions were mainly from industry and from the residential and commercial sector (Figure 6.72).

Spain

Figure 6.73. **Proportion of CO_2 emissions from energy use subject to different levels of effective carbon rates in Spain in 2012**

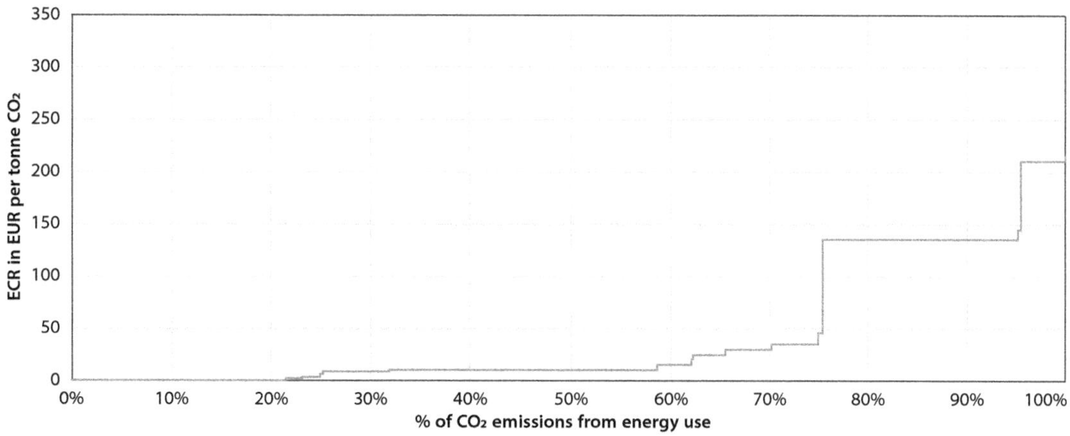

StatLink http://dx.doi.org/10.1787/888933390576

Figure 6.74. **Average effective carbon rates in Spain by sector and component in 2012**

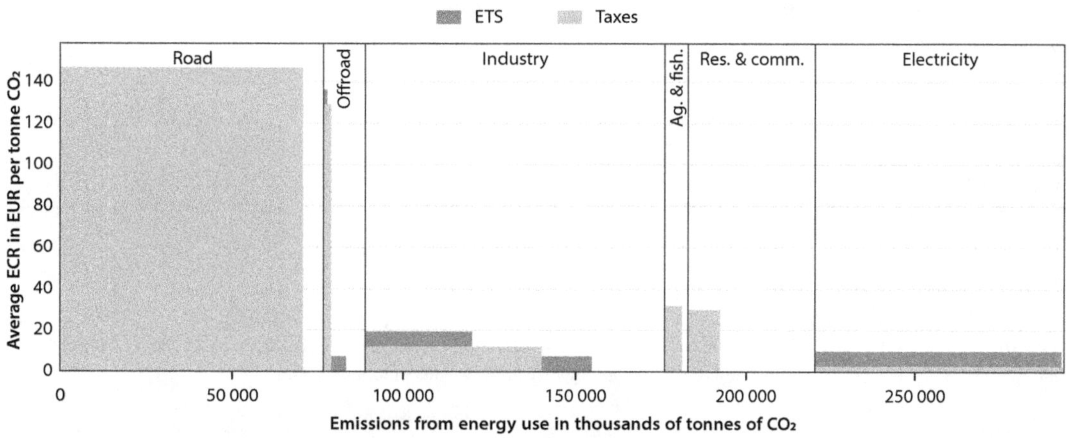

StatLink http://dx.doi.org/10.1787/888933390993

Note: See Box 3.1 in Chapter 3 for notes on the interpretation of effective carbon rates. Please see Annex A and Annex B for detail on methodology and underlying sources.

In 2012, effective carbon rates in Spain consisted primarily of specific taxes on energy use and to a very small extent of permit prices from the EU ETS. Spain did not have an explicit carbon tax. Spain priced 79% of carbon emissions from energy use, and 30% were priced above EUR 30 per tonne of CO_2 (Figure 6.73). The majority of these emissions were from the road sector. Carbon prices from tradable permits were a significant component of the effective carbon rate in the electricity sector. Unpriced emissions were primarily from industry, as well as from the residential and commercial sector (Figure 6.74).

Sweden

Figure 6.75. **Proportion of CO_2 emissions from energy use subject to different levels of effective carbon rates in Sweden in 2012**

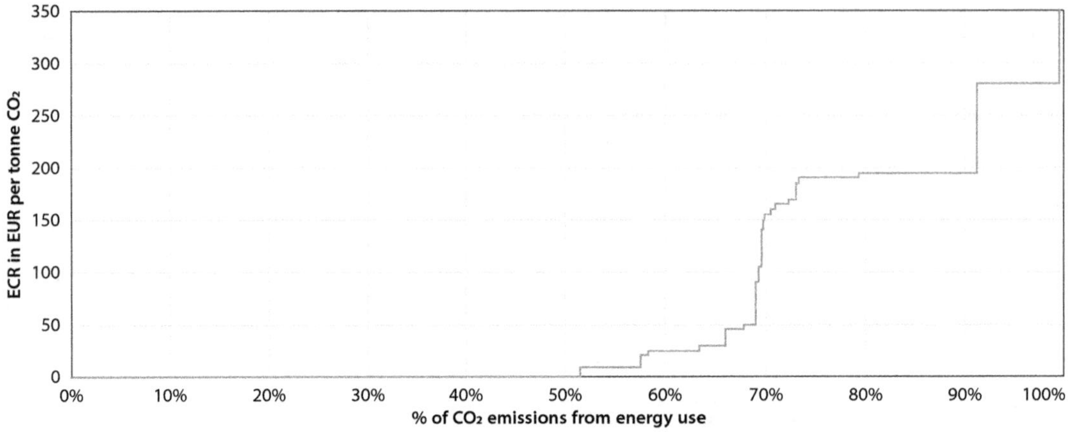

StatLink http://dx.doi.org/10.1787/888933390588

Figure 6.76. **Average effective carbon rates in Sweden by sector and component in 2012**

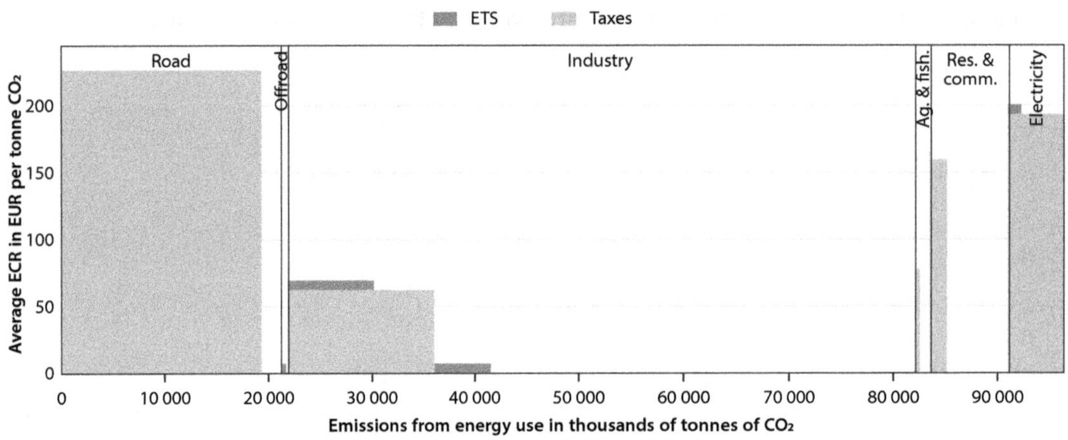

StatLink http://dx.doi.org/10.1787/888933391004

Note: See Box 3.1 in Chapter 3 for notes on the interpretation of effective carbon rates. Please see Annex A and Annex B for detail on methodology and underlying sources.

In 2012, effective carbon rates in Sweden consisted of specific taxes on energy use, national carbon taxes and permit prices from the EU ETS. Sweden priced 49% of carbon emissions from energy use, and 34% were priced above EUR 30 per tonne of CO_2 (Figure 6.75); the majority of these emissions were from the road sector. Unpriced emissions were primarily from industry and the residential and commercial sector (Figure 6.76).

A large share of these unpriced emissions was from the combustion of biomass. When excluding emissions from biomass (i.e. taking a lifecycle approach when accounting for the emissions from biomass), 90% of CO_2 emissions from energy use in Sweden were priced at a positive ECR, and 60% were priced above EUR 30 per tonne of CO_2. In the residential and commercial sector, 99% of total CO_2 emissions were priced at an ECR above EUR 30 per tonne of CO_2 when excluding emissions from biomass. In the industry sector, 86% of emissions were priced when excluding emissions from biomass, and 15% were priced above EUR 30 per tonne of CO_2.

Switzerland

Figure 6.77. **Proportion of CO_2 emissions from energy use subject to different levels of effective carbon rates in Switzerland in 2012**

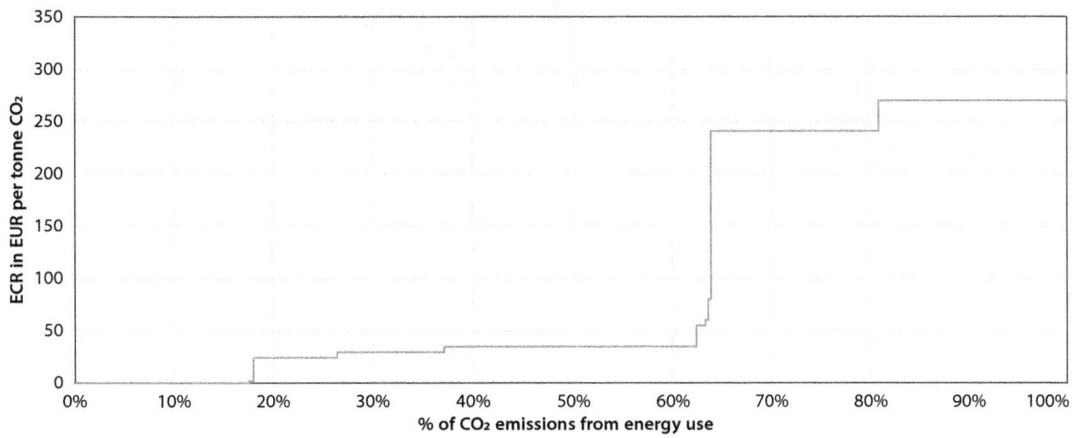

StatLink http://dx.doi.org/10.1787/888933390596

Figure 6.78. **Average effective carbon rates in Switzerland by sector and component in 2012**

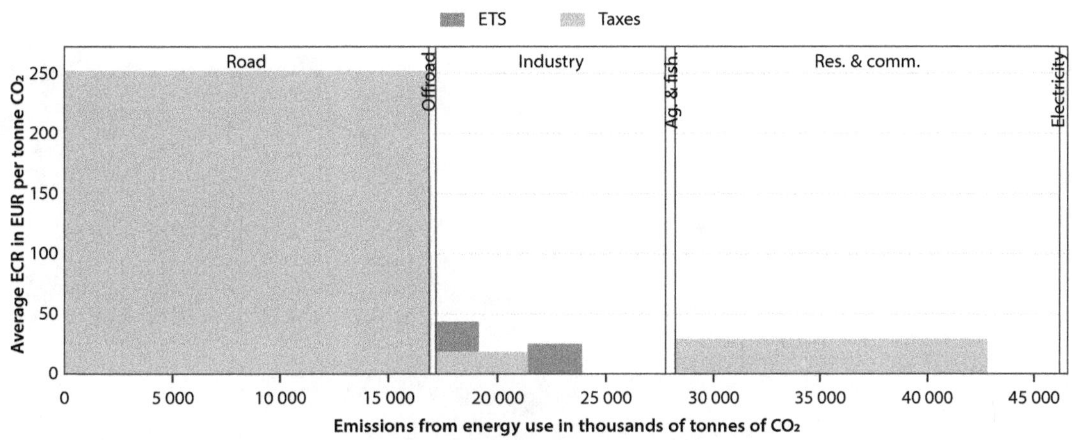

StatLink http://dx.doi.org/10.1787/888933391019

Note: See Box 3.1 in Chapter 3 for notes on the interpretation of effective carbon rates. Please see Annex A and Annex B for detail on methodology and underlying sources.

In 2012, effective carbon rates in Switzerland consisted primarily of specific taxes on energy use and to a smaller extent of national carbon taxes and permit prices from the Swiss ETS.[4] Switzerland priced 82% of carbon emissions from energy use, and 63% were priced above EUR 30 per tonne of CO_2 (Figure 6.77); a large share of these emissions was from the road sector. The Swiss ETS applied mainly to emissions of the industry sector. In that sector, 24% of emissions were covered by ETS only, 21% of emissions were covered by tax only, and 19% of emissions were covered by both taxes and the ETS. Unpriced emissions were primarily from industry and the residential and commercial sector (Figure 6.78).

The overlap between instruments occurred within the industry sector, between the mineral oils duty (not the mineral oil surtax, as this applied only in transport) and the emissions trading system. This is since the mineral oils tax also applied to fuels used for heating and process use, consistent with the information included in *Taxing Energy Use: A Graphical Analysis* (OECD, 2013, and more specifically the table shown on page 213 within that publication).

The high effective carbon rate on road transport energy is largely the result of mineral oil taxes, the revenues of which are largely earmarked to road infrastructure funding. These taxes are included because the tax base, mineral oils, is part of the tax base considered in the effective carbon rates. Other countries may fund road infrastructure via road tolls, which are not included in the effective carbon rate because road use is not part of the tax base considered in the effective carbon rates.

Turkey

Figure 6.79. **Proportion of CO_2 emissions from energy use subject to different levels of effective carbon rates in Turkey in 2012**

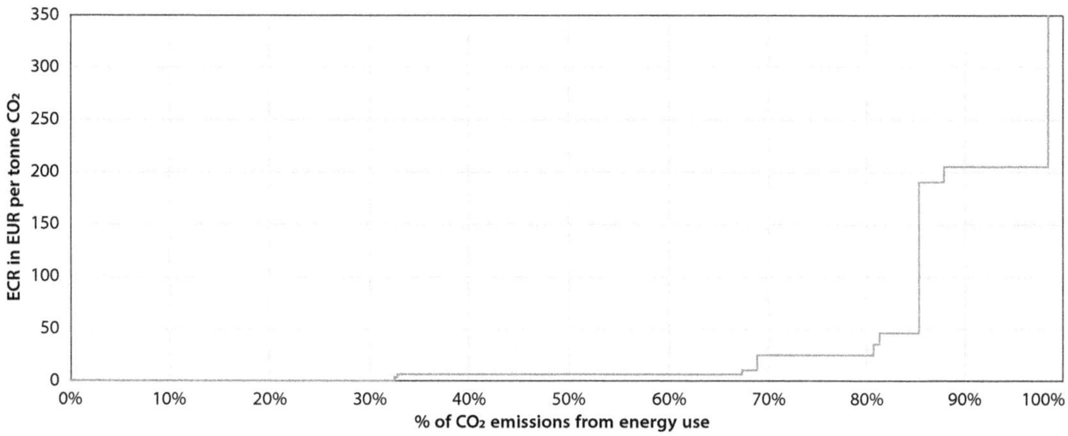

StatLink http://dx.doi.org/10.1787/888933390603

Figure 6.80. **Average effective carbon rates in Turkey by sector and component in 2012**

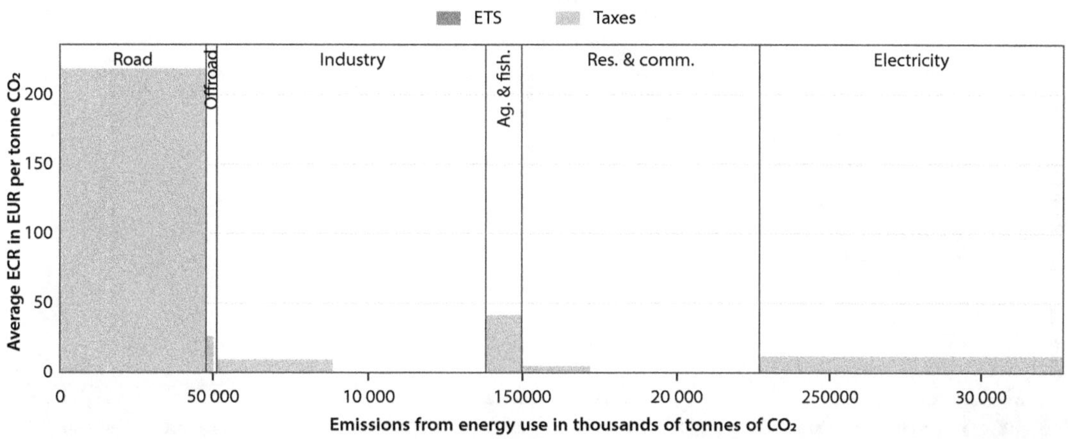

StatLink http://dx.doi.org/10.1787/888933391025

Note: See Box 3.1 in Chapter 3 for notes on the interpretation of effective carbon rates. Please see Annex A and Annex B for detail on methodology and underlying sources.

In 2012, effective carbon rates in Turkey consisted entirely of specific taxes on energy use. Turkey did not have an explicit carbon tax or an emissions trading system. Turkey priced 68% of carbon emissions from energy use, and 19% were priced above EUR 30 per tonne of CO_2 (Figure 6.79); the majority of these emissions were from the road sector. The majority of unpriced emissions were from industry and the residential and commercial sectors (Figure 6.80).

United Kingdom

Figure 6.81. **Proportion of CO_2 emissions from energy use subject to different levels of effective carbon rates in the United Kingdom in 2012**

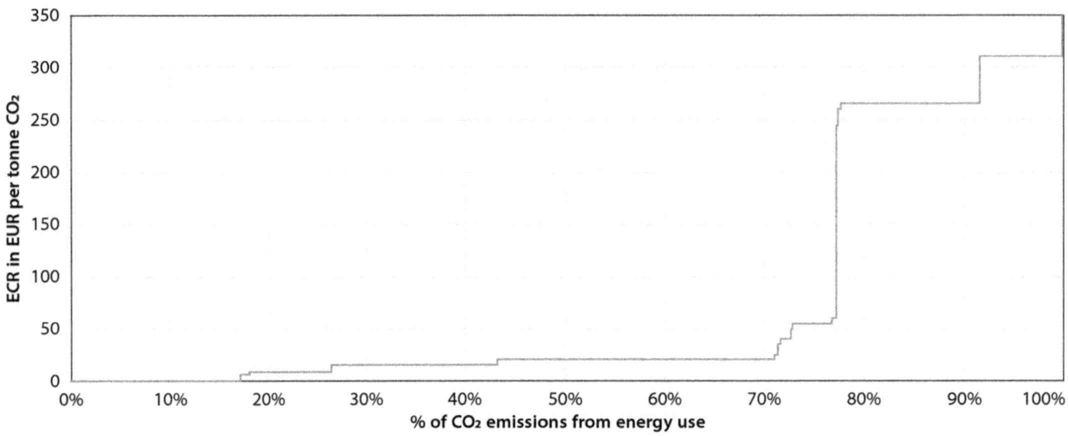

StatLink http://dx.doi.org/10.1787/888933390611

Figure 6.82. **Average effective carbon rates in the United Kingdom by sector and component in 2012**

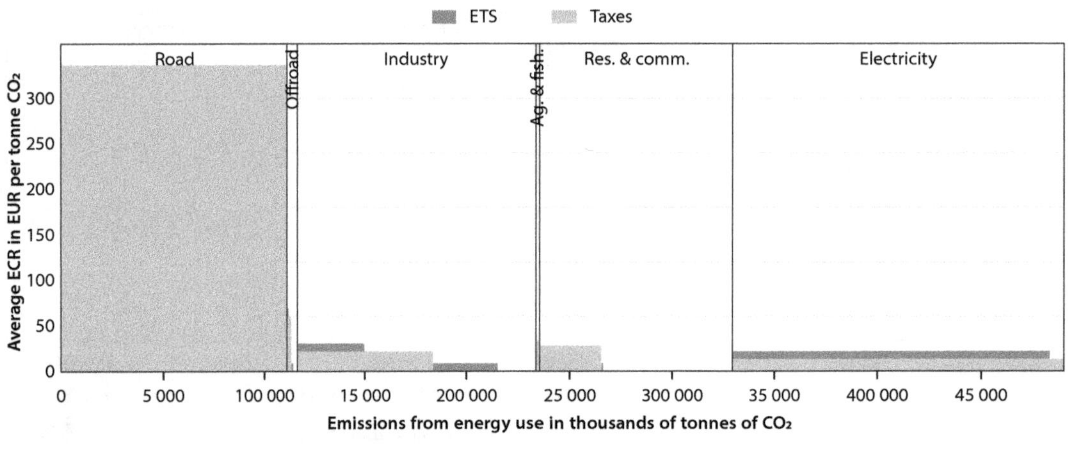

StatLink http://dx.doi.org/10.1787/888933391034

Note: See Box 3.1 in Chapter 3 for notes on the interpretation of effective carbon rates. Please see Annex A and Annex B for detail on methodology and underlying sources.

In 2012, effective carbon rates in the United Kingdom consisted primarily of specific taxes on energy use and to a smaller extent of national carbon taxes and permit prices from the EU ETS. The United Kingdom priced 83% of its energy related CO_2 emissions, and 29% of emissions were priced above EUR 30 per tonne of CO_2 (Figure 6.81); a large share of these emissions was from the road sector. Within the industry sector, the overlap between taxes and prices from tradable permits was just below 30%, while taxes and permit prices alone covered roughly the same share of emissions each. The overlap between tax and ETS coverage was almost 100% in the electricity sector. Unpriced emissions were mainly from industry and the residential and commercial sectors (Figure 6.82).

United States

Figure 6.83. **Proportion of CO_2 emissions from energy use subject to different levels of effective carbon rates in the United States in 2012**

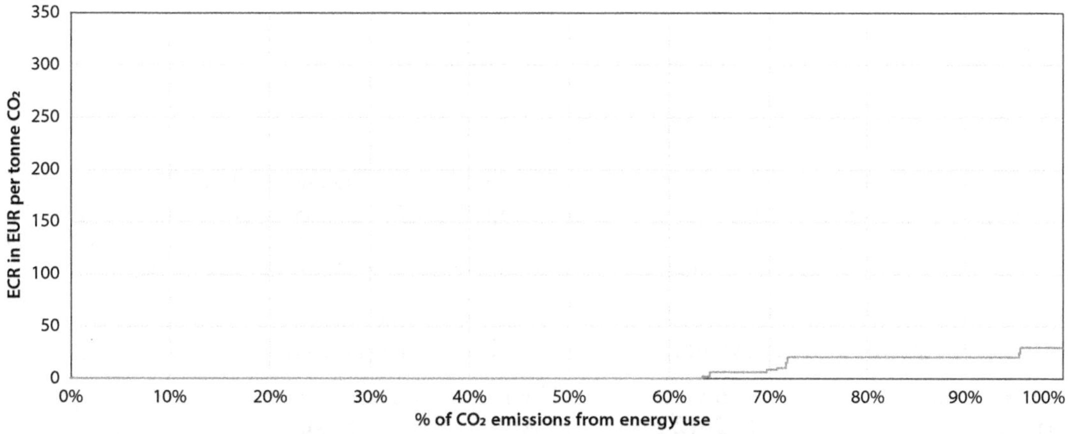

StatLink http://dx.doi.org/1787/888933390626

Figure 6.84. **Average effective carbon rates in the United States by sector and component in 2012**

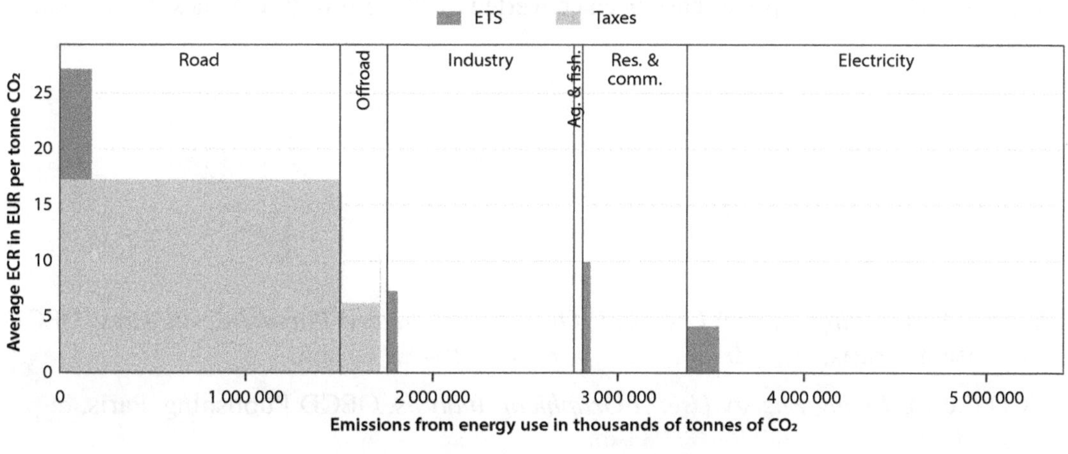

StatLink http://dx.doi.org/10.1787/888933391049

Note: See Box 3.1 in Chapter 3 for notes on the interpretation of effective carbon rates. Please see Annex A and Annex B for detail on methodology and underlying sources.

In 2012, effective carbon rates in the USA consisted primarily of specific federal taxes on energy use. Other pricing instruments were applied at the subnational level:[5] California has implemented an ETS and nine North-East Atlantic states take part in the Regional Greenhouse Gas Initiative (RGGI). The United States priced 37% of carbon emissions from energy use, none of which were priced above EUR 30 per tonne of CO_2 (Figure 6.83). The majority of priced emissions were from the road sector. The California Cap-and-Trade Program and RGGI covered emissions in the industrial, the residential and commercial, and the electricity sectors, and the price signal from tradable emissions permits entirely overlapped with that of federal taxes in the road sector. In total, both emissions trading systems together covered roughly 8% of CO_2 emissions from energy use in the USA. The majority of unpriced emissions were from industry, the residential and commercial sector and the electricity sector (Figure 6.84).

Notes

1. All taxes on energy use are shown for 2012. The Quebec ETS is shown as in 2013 (including the extension to fossil fuel suppliers in 2015), and permit prices have been converted to 2012 Euros (see Annex A for details). The carbon tax in British Columbia is not shown in OECD (2013b), but has been included for the estimation of effective carbon rates as in 2012.
2. All taxes on energy use are shown for 2012. The seven subnational Chinese ETS are shown as in 2014, and permit prices have been converted to 2012 Euros (see Annex A for details).
3. All taxes on energy use are shown for 2012. The Korean ETS is shown as in 2015, and permit prices have been converted to 2012 Euros (see Annex A for details).
4. All taxes on energy use are shown for 2012. The Swiss ETS is shown as in 2014, and permit prices have been converted to 2012 Euros (see Annex A for details).
5. All taxes on energy are shown for 2012. The California ETS is shown as in 2013 (including the extension to fossil fuel suppliers in 2015), and the Regional Greenhouse Gas Initiative is shown as in 2012. Permit prices have been converted to 2012 Euros (see Annex A for details).

References

OECD (2015), *Taxing Energy Use 2015: OECD and Selected Partner Economies*, OECD Publishing, Paris, http://dx.doi.org/10.1787/9789264232334-en.

OECD (2013), *Taxing Energy Use: A Graphical Analysis*, OECD Publishing, Paris, http://dx.doi.org/10.1787/9789264183933-en.

Annex A

Estimating effective carbon rates

This Annex describes the methodology used to estimate the effective carbon rates (ECRs) presented in this report. Section A.1 provides details on the concept of ECRs and their three components: carbon taxes, specific taxes on energy use and tradable emission permit prices. Section A.2 gives an overview of the data and methodology used to estimate permit prices and coverage of emissions trading systems. Section A.3 discusses how the estimates of permit prices and ETS coverage are combined with information on carbon taxes and other specific taxes on energy use to produce estimates of ECRs.

Annex B describes the sixteen emissions trading systems analysed for the purposes of this report. It also shows results on average effective ETS prices and ETS coverage, and explains the ETS-specific adjustments and assumptions that were made in the calculations of ECRs.

A.1. Effective carbon rates: Concept and scope

The effective carbon rate (ECR) is the total carbon price that applies to CO_2 emissions from energy use as a result of market-based policy instruments. The ECR adds up taxes and tradable emission permit prices and has three components:

1. carbon taxes, which set a tax rate on the carbon content of each form of energy,
2. other specific taxes (primarily excise taxes) on energy use, which are typically set per physical unit or unit of energy, but which can be translated into effective tax rates on the carbon content of that form of energy, and
3. the price of tradable emission permits that must be surrendered per unit of CO_2 emissions, regardless of how it was acquired, representing the opportunity cost of emitting an extra unit of CO_2.

At the most disaggregated level, the ECRs discussed in this report are estimated separately for 30 different users, 61 different fuels and a binary distinction of whether emissions from these users and fuels are covered by an ETS or not. In total, the combination of carbon taxes, other specific taxes on energy use and tradable emission permit prices therefore is assessed 3660 times. The components can, and often do, equal zero, depending on the user and fuel. They vary widely across users and fuels within a jurisdiction, as well as between jurisdictions.

By way of example, Figure A.1 illustrates the three components of the effective carbon rate for two types of users (households and paper production) and three fuels (diesel, natural gas and coal). In this hypothetical example, diesel use by households is subject to all three components of the effective carbon rate (a situation not found in current practice anywhere), tradable emission permit prices, carbon taxes and other specific taxes on energy use. However, natural gas use by households is not subject to an emissions trading system (ETS) and hence the associated effective carbon rate does not include a tradable emission permit price component. Coal use by households is not subject to any tradable emission permit price nor to any taxes and thus the associated effective carbon rate does not include a permit price nor any tax components. Paper production, which is taken as an example of an energy-intensive industry, is subject to one or two components of the ECRs in this hypothetical example. While for coal use it is only subject to tradable emission permit prices, diesel and natural gas use have an additional specific energy tax component, though at different effective rates per tonne of CO_2.

Figure A.1. **Illustration of composition of effective carbon rates**

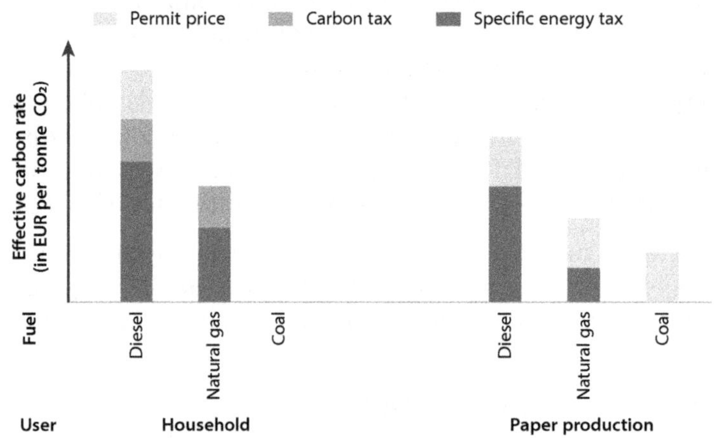

The rest of the section discusses the three components of the ECRs separately. The data and detailed methodology used to calculate ECR coverage are discussed in Sections A.2 and A.3.

A.1.1. Carbon taxes

Carbon tax rates are usually set for each fuel on the basis of their CO_2 content. In case there is a fixed rate per unit of CO_2, the result is a uniform carbon tax, a desirable feature from a cost-effectiveness point of view. For administrative purposes, for each fuel the tax can be translated into tax rates per litre, kilogram, or gigajoule of energy depending on the carbon content of the fuel. Some carbon taxes may specify different rates for different fuels or users even in carbon terms.

Carbon tax rates for different users and fuels as at 1 April 2012 are reported in the *Taxing Energy Use* (TEU) database. Several OECD countries, including Denmark, Finland, Iceland, Ireland, Slovenia, Sweden and Switzerland, applied carbon taxes to certain forms of energy use at that date. Except for the carbon tax in British Columbia, subnational carbon taxes are not included in the analysis.

Carbon taxes are often entirely or partially alleviated if the energy user is subject to an ETS. The TEU database does not include information on the coverage of emissions trading systems and consequently does not account for the reduction or removal of carbon taxes when an ETS applies. To calculate the impact of carbon tax rates in ECRs, and to ensure that the interactions between ETS and carbon taxes are taken into account, it is necessary to adapt the information on carbon taxes in the TEU database (see Section A.3.3). Since interactions always imply lower or equal, but never higher, taxes on energy use that is subject to an ETS, the price signal from taxes shown in this report is slightly lower for some countries than the tax rate shown in the TEU reports (OECD, 2013 and 2015).

A.1.2. Other specific taxes on energy use

All the OECD and G20 countries included in the TEU database levy taxes on certain forms of energy use, often also, or primarily, for the purpose of raising revenue. These taxes, most commonly excise taxes, are the most significant component of ECRs. (Value-added tax is not included since it is usually not specific to energy. Transport tolls are excluded because they are not directly related to energy use but instead relate to infrastructure use.) Taxes on transport fuels apply in all countries. Their rates are generally higher than those on other fuel use. Heating and process fuels, fuels used as inputs to electricity generation as well as electricity consumption are less commonly taxed. If they are taxed, the rates applied are generally lower than those on fuels used in transport (OECD, 2013 and 2015).

These other specific taxes on energy use are typically levied in terms of physical units (e.g. litres, kilograms, m^3), or in terms of energy content (e.g. GJ or kWh), and not by reference to the carbon content of the fuel. However, these taxes can be translated into effective tax rates on the carbon content of the fuel due to the proportional relationship between these characteristics and the carbon content of each type of fuel. In that sense, in terms of changing fuel prices and thereby affecting the behaviour of fuel users, they are equivalent to carbon taxes in putting a price on carbon, albeit less consistent from an environmental perspective in the approach to the rates applied. For example, a tax set uniformly on the carbon content of a unit of energy from bituminous coal would be around 1.75 times as high as a tax on the carbon content on a unit of energy from natural gas, whereas a tax on energy content would be the same per unit of energy. Taxing the carbon content at the same rate for both fuels would increase the price of both forms of energy use, but would constitute a stronger signal to move to natural gas than would a tax on energy content, as natural gas would be relatively cheaper than bituminous coal under the tax based on carbon content.

The information regarding tax rates on other specific taxes on energy use by user and fuel is as at 1 April 2012 and is obtained from the updated TEU database.[1]

A.1.3. Prices of tradable emission permits

The third component of ECRs is the tradable emission permit price resulting from emissions trading systems. This report covers permit prices from emissions trading systems in the 41 countries in the TEU database, encompassing supra-national, national and subnational jurisdictions. Sixteen emissions trading systems have been included in the analysis: the Beijing Emissions Trading System (China), the California Cap-and-Trade Program (United States), the Chongqing Emissions Trading System (China), the European Union Emissions Trading System (which operates in 31 countries, of which 23 are OECD member countries), the Guangdong Emissions Trading System (China), the Hubei Emissions Trading System (China), the Korea Emissions Trading Scheme, the New Zealand Emissions Trading Scheme, the Quebec Cap-and-Trade System (Canada), the Regional Greenhouse Gas Initiative (RGGI, covering nine north-east and mid-Atlantic US states), the Saitama Prefecture Target Setting Emissions Trading System (Japan), the Shanghai Emissions Trading System (China), the Shenzhen Emissions Trading System (China), the Swiss Emissions Trading Scheme, the Tianjin Emissions Trading System (China) and the Tokyo Cap-and-Trade Programme (Japan).

Average permit prices at auctions are calculated across a year, if the data is available. An average is taken to smooth price fluctuations, where possible. For some emissions trading systems, price information is only available for part of the year, in which case an average across the available dates is calculated, or for a single auction, in which case this auction price is used. Due to data availability, secondary market prices rather than auction prices are used in the calculation for the Chinese subnational systems and the emissions trading systems in Japan, Korea and New Zealand.

Permit prices are for 2012 for emissions trading systems that started operating before this date, in order to be consistent with the tax rate information from the TEU database. Where emissions trading systems became operational after 2012, permit prices for the first year of operation[2] are used after adjusting for inflation back to 2012 price levels. Consequently, while all tax rates are for 2012, permit prices are for different years. This effectively assumes that supply and demand of permits would have resulted in the same permit price, had the ETS existed in 2012. Section A.2.1 provides further detail on permit prices for each ETS.

To incorporate permit prices from emissions trading systems into ECRs, it is also necessary to estimate the coverage of the ETS, i.e. the share of emissions that are subject to an ETS. While tax coverage is directly determined by the users and fuels that are subject to the tax, ETS coverage is an estimate as it applies to emissions of a facility subject to an ETS and does not distinguish between fuels, and facility level data need to be matched to energy balance data. For most systems, ETS coverage is estimated by reference to verified emissions data at facility level or at aggregated facility level. Where this is not available, jurisdiction-specific information on fuel use or on permit allocation is used. Section A.2 gives further detail on how the ETS coverage was estimated.

ETS coverage is estimated for six sectors, which together comprise all CO_2 emissions from energy use in each country. These sector definitions can differ from those used in individual countries. The sectors are:

- road transport – all energy used in road transport,
- off road transport – all energy used in offroad transport, including pipelines, rail transport, domestic aviation and maritime transport,
- industry – all energy used in industrial processes, in heating (including inside industrial facilities) in the transformation of energy and for auto-generation of electricity in industrial facilities,
- agriculture and fisheries – all energy used in agriculture, fisheries and forestry except for on-road transport (which is included in road transport),
- residential and commercial – all energy used for commercial and residential heating, and
- main electricity generation – all fuels used in main electricity generation, so not including auto-generation in industrial facilities.

A.2. Data and methodology to calculate permit prices and estimate ETS coverage

To estimate the combined impact of taxes and trading systems in putting a price on CO_2 emissions from energy use, permit prices and coverage of emissions trading systems must be integrated with the information on tax rates and coverage in the TEU database. A necessary step is to estimate to what extent the different users of energy in the TEU database are subject to each ETS, and at what price. To do this, the data for each ETS must be consistent with the TEU database in terms of the sector groupings and the emissions covered.

This section details how permit prices and the coverage of the emissions trading systems are estimated, describing first the general approach in Section A.2.1 and then looking at the detailed estimation strategy necessary to match information on ETS coverage from diverse data sources with the TEU database in Section A.2.2.

A.2.1. General approach for calculating permit prices and estimating ETS coverage

This section first discusses how permit prices are calculated and second how ETS coverage is estimated.

A.2.1.1. Calculation of permit prices

The price signal from emissions trading systems on CO_2 emissions from energy use is calculated, where information is available, by taking the average permit price observed at auction across a year of operation. Where possible, an average price across a year of operation is used to smooth the effects of short-term fluctuations in permit prices. The average permit price is assigned to all CO_2 emissions from energy use subject to an ETS,[3] regardless of whether an emitter surrendered a permit that was purchased or obtained for free. This treats the permit price as the opportunity cost of all emissions and permits.

Where more than one ETS operates in the same country (California and RGGI in the United States, and Beijing, Chongqing, Guangdong, Hubei, Tianjin, Shanghai and Shenzhen, in China), an average permit price from all emissions trading systems within that country is constructed for each of the six sectors, weighted by the coverage of each ETS within that sector.[4]

For emissions trading systems that were operational in 2012 (EU, Japan, RGGI and New Zealand), the prices of permits in 2012 are used. For emissions trading systems that came into operation after 2012, an average for their first year of operation is used except for Switzerland, where no auctions were held in the first year of operation, and for the subnational emissions trading systems in China, where 2014 is used for all emissions trading systems. Where a later year is used, permit prices are adjusted to 2012 prices using the growth rate of consumer price indices for each country. They are converted to EUR using the conversion rates for each currency used in the TEU database.

For some emissions trading systems, auction data is not available or is available for less than a full year. For the emissions trading systems in China, the permit price is calculated based on secondary market data across the longest period available in 2014 (part years for Beijing, Guangdong Shanghai, Shenzhen, Tianjin and Hubei, and a single point estimate for Chongqing). In the ETS operating in Québec, only one auction has taken place in the year for which emissions data is available, so prices at this auction are used. In Japan, the midpoint of an estimated range of the cost of one excess reduction credit by the Tokyo Municipal Government is used. For the Korean ETS, the average permit price as traded at the Korea Exchange in the first half of 2015 is used, since no auctions of permits have taken place. In New Zealand, half of the average price of Joint Implementation Credits in 2012 is used[5] to reflect the fact that 79% of surrendered permits in 2012 were Joint Implementation Credits (New Zealand Ministry for the Environment, 2012).

Table A.1 shows the average permit price for the different emissions trading systems, in local currency and converted into EUR as at 2012.

Table A.1. **Average ETS price signals used in the report**

ETS	Price signal (local currency)	Currency	Year used	Price signal (2012 EUR)
EU	7.24	EUR	2012	7.24
New Zealand	2.18	NZD	2012	1.33
RGGI	1.93	USD	2012	1.48
Saitama and Tokyo	9 450.00	JPY	2012	92.04
California	12.83	USD	2013	9.91
Québec	10.75	CAD	2013	8.05
Beijing	55.00	CNY	2014 (partial)	6.66
Chongqing	30.70	CNY	2014 (partial)	3.72
Guangdong	62.10	CNY	2014 (partial)	7.52
Hubei	23.70	CNY	2014 (partial)	2.87
Shanghai	36.20	CNY	2014 (partial)	4.38
Shenzhen	73.50	CNY	2014 (partial)	8.9
Tianjin	30.90	CNY	2014 (partial)	3.74
Switzerland	30.13	CHF	2014	24.97
Korea	10 177.00	KRW	2015 (partial)	6.66

A.2.1.2. Estimation of ETS coverage

The coverage of each ETS is estimated as a share of total emissions, by dividing the CO_2 emissions from energy use that are subject to an ETS (covered emissions) by the total amount of CO_2 emitted. This is done for each of the six sectors listed in Section A.1.3. Where subnational emissions trading systems apply, the emissions subject to the ETS are divided by the total amount of CO_2 emissions in the country, so that the estimates of ETS coverage apply at the national level, consistent with the tax information. More specifically, the sector coverage of an ETS is given by:

$$ETS\ coverage_{i,c,y} = \frac{\sum_j emissions\ subject\ to\ ETS_{j,i,r,y}}{total\ emissions_{i,c,2012}} \qquad (1)$$

Where:

j is the facility that is subject to the relevant ETS regulation;

i is the sector in which facility j is active (road, offroad, industry, residential and commercial, agriculture and fisheries, main electricity generation);

r is the (sub)national jurisdiction to which the ETS regulation applies;

c is the country in which (sub)national jurisdiction r is embedded;

y is the year of analysis.

Covered emissions in each sector, the numerator of equation (1), are estimated as described in Section A.2.2. Information on total emissions in each sector in each country, the denominator of equation (1), is taken from the TEU database,[6] with some adjustments described in Section A.3.1 and A.3.2. This information is for 2012.

A.2.2. Detailed estimation strategy: Calculating CO_2 emissions subject to emissions trading systems

Calculating the total emissions from each sector subject to each ETS (covered emissions), the numerator of equation (1), requires coherent definitions and a correct sector allocation of emissions based on different data sources. Reconciling these data sources and ensuring consistency with the TEU database on CO_2 emissions from energy use, the denominator of equation (1), requires a number of estimations. The remainder of this section discusses the detailed estimation strategy undertaken to allocate ETS-covered emissions to different sectors and to ensure consistency with the TEU database.

A.2.2.1. Data sources for calculation of covered emissions (numerator)

Data on the CO_2 emissions from energy use subject to the ETS in each jurisdiction, are from national sources. Where possible, verified emissions data from the authorities responsible for administering the ETS is used. For the EU, Switzerland, RGGI, California and Quebec systems, data on verified emissions at facility level has been obtained (European Commission 2014a, European Commission 2014b, Swiss Emissions trading Registry 2015, RGGI 2015, California Environmental Protection Agency Air Resources Board 2014, Gouvernement du Québec 2013). The data includes the actual emissions per year (in tonnes of CO_2) of all facilities that are subject to the ETS. The facilities are matched to the individual industries in the TEU database using industry codes provided with the verified emissions data (NACE for the EU and NAICS for California, Quebec and RGGI) and then subsequently grouped into the six sectors listed in Section A.1.3.[7] For the Swiss ETS, information provided by national officials (Swiss Federal Office for the Environment, 2015) is used to match facilities' emissions data to the individual industries in *Taxing Energy Use*. In the Saitama and Tokyo emissions trading systems and the seven Chinese emissions trading systems, verified emissions data is available at an aggregated facility level (Saitama Prefectural Government 2014, Tokyo Metropolitan Government 2012, Li et al. 2015). This aggregated facility data could be grouped into the six sectors listed in Section A.1.3.

Other approaches were followed due to limitations in data availability for some emissions trading systems or due to features of an ETS which made facility level data inappropriate:

- The ETS in Korea came into force in 2015 and data on verified emissions is not yet available. In this case, data on permit allocations to individual sectors of the economy, taken from government websites, is used to estimate ETS coverage (Korean Ministry of Environment, 2014).
- The New Zealand ETS applies to upstream emissions, covering imports and production of liquid fuels, coal and natural gas. Data on the total level of emissions from each fuel subject to the ETS is applied to the total supply of each fuel for domestic use to determine the proportion of energy consumption by each sector (New Zealand Environmental Protection Authority, 2012).

Annex B contains further information on the data sources for each jurisdiction.

A.2.2.2. Auto-generation of electricity and heat

Electricity and heat can be produced as a main activity (main generation) or by (mainly) industrial or energy transformation users that also produce electricity (auto-generation).

The TEU database provides a single entry (disaggregated by fuel) that includes all emissions from auto-generation of electricity, and another that includes all emissions from the auto-generation of heat, regardless of the industry in which the auto-generation was undertaken. In the TEU publications, the emissions from the auto-generation of heat and electricity are shown with main generation of electricity and heat respectively.

Emissions trading systems, however, are operated at a facility level. If a facility exceeds the threshold for inclusion into an ETS, emission permits must be submitted for all emissions from that facility. Since large facilities subject to emissions trading systems (e.g. petroleum processing or manufacturing plants) may generate their own electricity and heat, such facilities will include emissions from the auto-generation of electricity or heat in the emissions which are reported to registries. Consequently, data on CO_2 emissions subject to emissions trading systems includes emissions from fuels used in the auto-generation of electricity or heat with the emissions from the relevant industry and does not separately identify them.

To calculate the coverage of ETS by sector, emissions from the auto-generation of heat and electricity must be treated consistently in the covered emissions (the numerator of equation (1), estimated from jurisdiction-specific data), with total emissions (the denominator of equation (1), taken from the TEU database). Ideally, this requires that in the jurisdiction-specific data, emissions from the auto-generation of electricity and heat in each facility or individual industry be identified separately from other industrial emissions in that sector. Once identified, these auto-generation emissions would be grouped into a single entry as auto-generation emissions for electricity or heat, respectively.

However, consistent and comprehensive data is not publicly available on the proportion of covered emissions from each facility or group of facilities that result from auto-generation, so separating these from other covered industrial emissions is not possible. Given that it is not possible to identify these emissions in the jurisdiction-specific information from which covered emissions are calculated (the numerator of equation (1)), the TEU database information on total emissions (the denominator of equation (1)) has been adjusted instead. In the TEU database, the total amount of emissions from the auto-generation of heat and electricity are separately identified, although the industries they pertain to are not identifiable. These have therefore been removed from the electricity sector and placed in the industrial sector. Effectively, both covered emissions and total emissions have been grouped at the lowest consistent level of aggregation – total industry emissions, including all auto-generation – between the data on covered emissions and the TEU database.[8]

The industry sector therefore includes industry emissions with emissions from auto-generation, removing the need to identify auto-generation emissions separately for each of the individual industries (in the jurisdiction-specific data) or to allocate auto-generation emissions to individual industries (in the TEU database). Section A.3.1 provides further information on changes to the sectors presented in the TEU database and publications.

A.2.2.3. Emissions from industrial processes (non-energy use industrial emissions)

All of the emissions trading systems considered include CO_2 emissions from non-energy sources, such as emissions from industrial processes. CO_2 emissions from industrial processes occur during the manufacturing of a limited set of products in several industries as a result of chemical and physical transformations, independent from energy use.

Allocation data or verified emission data from facilities include all CO_2 emissions from these facilities, whether arising from energy combustion or from industrial processes. For consistency with the TEU database, which reports only emissions from energy combustion and not those resulting from industrial processes, it is necessary to identify and exclude these industrial process emissions when calculating covered emissions (the numerator of equation (1)).

Data from the UNFCCC National Inventory Reports (UNFCCC, 2015) is used to identify industrial process emissions. The UNFCCC reports show total (sub)national emissions from all sectors and differentiate between CO_2 emissions from different sources (including energy-based and process-based emissions). The industrial process emissions for individual industries (typically the chemical, iron and steel, non-ferrous metals, and non-metallic industries) or for the whole industry sector are calculated from the UNFCCC data. These emissions are deducted from the (sub)national data on verified emissions to isolate the CO_2 emissions from energy use. Annex B provides additional details for each of the emissions trading systems.[9]

This approach implicitly assumes that industrial process emissions are uniformly distributed across the facilities that are and are not subject to an ETS. However, facilities that are subject to the ETS are generally the large and heavy emitting firms with a relatively higher share of industrial process emissions. As the UNFCCC (2015) inventories report industrial process emissions of both large and small firms, a share relying on this data is likely to underestimate the industrial process emissions of firms subject to the ETS. Therefore, since a likely underestimate is subtracted from total emissions under an ETS, the estimate of ETS coverage of energy use alone in the industry sector is likely to be an overestimation.

A.2.2.4. Emissions of non-CO_2 greenhouse gases

Many of the emissions trading systems cover other greenhouse gas emissions from energy use, such as methane (CH_4), nitrous oxide (N_2O), sulphur hexafluoride (SF_6), hydrofluorocarbons (HFCs), perfluorocarbons (PFCs), and nitrogen trifluoride (NF_3). Emissions trading systems that cover greenhouse gases other than CO_2 in the year of analysis include those operating in California, Chongqing, Korea, New Zealand, Québec and Switzerland.[10] Emissions trading systems covering only emissions of CO_2 include Beijing, the EU (prior to 2013), Guangdong, Hubei, RGGI, Saitama, Shanghai, Shenzhen, Tianjin and Tokyo.

In the emissions trading systems where other greenhouse gases are included, these need to be excluded from emissions of CO_2 to calculate the total emissions from each sector subject to each ETS (the numerator of equation (1)). For each of these systems except New Zealand,[11] information provided in the UNFCCC (2015) reports[12] is used to identify CO_2 emissions. These reports detail national emissions by activity and differentiate between the different greenhouse gases. The share of CO_2 emissions to total greenhouse gas emissions, measured as CO_2 equivalent emissions, is used to isolate CO_2 from the total greenhouse gas emissions subject to each of these emissions trading systems.

A.2.2.5. Biomass

Some emissions trading systems treat CO_2 emissions from the combustion of biomass differently than CO_2 emissions from other sources, due to differences in approaches to accounting for their net lifecycle emissions. In several emissions trading systems, CO_2 emissions from the combustion of biomass are not included. In the remaining systems, emissions from biomass are subject to the ETS when emitted by a facility subject to the ETS, but are zero-rated when calculating covered emissions, meaning that no permit needs be surrendered for biomass emissions.

Among the emissions trading systems considered, biomass is not covered by the emissions trading systems in California, New Zealand, Québec, RGGI, Saitama, or Tokyo. In these cases, the emissions from each sector subject to each ETS (the numerator of equation (1)) do not include emissions from biomass. CO_2 emissions from biomass are included, but an emissions factor of zero is applied to their use, in the EU ETS and in the emissions trading systems in Korea, Switzerland and the regional emissions trading systems in China. In these emissions trading systems, biomass emissions are not included in the data on verified emissions or allocation permits, as no permits have to be surrendered in respect of them. However, as they are included in the ETS coverage in these jurisdictions, it is assumed that emissions from biomass are covered if the facility using the biomass is covered. When results are calculated inclusive of biomass, the resulting estimate of biomass coverage is included in the total emissions from each sector subject to each ETS (the numerator of equation (1)) but no permit price has been assigned to these emissions.

A separate issue is how biomass should be treated in the analysis of ECRs. Emissions from the combustion of biomass can be seen to differ from emissions from the combustion of fossil fuels in two ways. First, biomass may absorb CO_2 from the atmosphere during its growing phase, before emitting CO_2 on combustion. Second, the life-cycle of biomass is considerably shorter than that of fossil fuels. There are three possible approaches to the calculation of CO_2 emissions from biomass, including:

- CO_2 emissions calculated according to the release of CO_2 on combustion, as for other fuels,
- a carbon neutral approach, which assumes that the lifecycle emissions of biomass are zero due to absorption during their life-span fully offsetting their emissions on combustion, and
- a lifecycle approach, which considers the source of the biomass, land use change, and whether it is replanted, to estimate the net lifecycle carbon emission associated with the report.

The first two approaches represent two extremes and are likely to be too simplistic. However, in the absence of detailed investigation of the lifecycle emissions of biomass, the TEU reports have taken the first approach. Many jurisdictions, as well as the UNFCCC, take the second approach to the treatment of biomass, considering it to be carbon neutral.

For completeness, this report includes results under both the first and second approach. Under the first approach, CO_2 emissions from biomass at combustion are included in the total emissions in each country (i.e. in the denominator of equation (1)). This is the approach most consistent with the TEU reports. Under the second approach, biomass is treated as carbon neutral, and CO_2 emissions from biomass are not included in total emissions. This approach is more consistent with country and UNFCCC measurements of CO_2 emissions. Results for the third approach would likely lie between the first and second approach.

A.3. Combining taxes and emissions trading systems to estimate effective carbon rates

To estimate ECRs, permit prices and estimated ETS coverage must be integrated with the information on tax rates and tax coverage found in the TEU database. By combining information on carbon taxes, other specific taxes on energy use and emissions trading systems, the combined coverage of the three instruments (ECR coverage) and their combined price impact can be assessed.

Estimating ECRs and ECR coverage requires several steps. First, the sector breakdown in the TEU database must be adjusted to match the six sectors for which ETS coverage is calculated. Second, the estimates of ETS coverage must be merged with information on tax coverage from the TEU database, so that emissions from energy use for each fuel and user can be divided between emissions covered by emissions trading systems and emissions not covered by emissions trading systems. Third, the impact of ETS coverage on carbon taxes must be identified where applicable in the TEU database. Finally, ECRs are estimated. The remainder of this section discusses each of these steps in more detail.

A.3.1. Adjusting emissions by sector for emissions from auto-generated electricity

As described in Section A.1.3, ETS coverage and permit prices have been estimated for six economic sectors that together span all energy use: road use, offroad use, industrial production, residential and commercial, agriculture and fishing, as well as electricity generation. These sectors differ slightly from those used in *Taxing Energy Use,* notably with regard to the inclusion of auto-generated electricity with industrial use rather than with electricity generation.[13] Consequently, the tax base divisions presented in previous TEU reports (road use, offroad use, residential and commercial, industrial and energy transformation, and electricity) have to be adjusted by moving auto-generation of electricity to the industrial sector.

Table A.2 provides details on the energy use included in the different sectors in this report on ECRs and in previous TEU reports.

Table A.2. **Categorisation of energy use for the estimation of effective carbon rates**

	TEU	ECR
Road	All energy used in road transport.	Like TEU
Offroad	All energy used in offroad transport (incl. pipelines, rail transport, domestic aviation and maritime transport).	Like TEU
Industry	All energy used in industrial processes, in heating (incl. inside industrial installations) and in the transformation of energy.	Like TEU, but also including fuels used for auto-generation of electricity in industrial installations (see Section A.2.2.2. for details).[1]
Commercial & residential	All energy used for commercial and residential heating.	Like TEU
Agriculture & fisheries	Energy used in agriculture, fisheries and forestry. Energy used in on-road transport in this sector is included in the road transport sector.	Like TEU
Electricity	All fuels used to generate electricity for domestic use (rather than the amount of energy generated from each fuel).	Like TEU, but fuels used in the auto-generation of electricity have been removed to industrial production (see Section A.2.2.2 for details).[2]

Notes: 1. In China, auto-generation of electricity is included in the main electricity sector, because publicly available information does not allow the amount of auto-generation of electricity in total electricity generation to be identified at the level of the seven pilot systems. In practical terms, auto-generation of electricity in China only plays a prominent role in aluminium production, which is mainly located outside the seven pilot systems. Hence, this deviation from the general approach does not affect results in practice.

2. For consistency with TEU, verified emissions from electricity production in the EU ETS have been adjusted with the share of all electricity production that is consumed domestically, from TEU.

A.3.2. Calculating the combined coverage of taxes and emissions trading systems

The ECR is a single metric which measures the combined impact of two types of market-based mechanisms – taxes (including both carbon taxes and other specific taxes on energy use) and tradable emission permit prices – in putting a price on carbon emissions. The calculation of the combined coverage of these instruments (ECR coverage) requires comprehensive information on the emissions that are subject to taxes and the emissions that are subject to an ETS. The combination of taxes or ETS affects the proportion of emissions from energy use that is subject to a price signal from either a tax or an ETS. This section describes how information on the coverage of taxes and emissions trading systems is integrated to develop a metric measuring their combined coverage and price levels.

How precisely the information on tax and ETS coverage can be combined to estimate ECR coverage depends on the relative levels of disaggregation in the TEU database and the data on emissions trading systems. The TEU database provides information on tax rates and coverage for a total of 30 different individual users, and within these users, for emissions from 61 different fuels. Data on the coverage of each ETS is based on facilities' total emissions and not differentiated by fuel. Therefore, matching the estimates of ETS coverage to the TEU database relies on the assumption that emissions from each fuel used by an individual user are equally subject to an ETS. For example, if 40% of emissions from a user are subject to an ETS, the analysis assumes that 40% of emissions from each of the fuels used by that user are subject to the ETS.[14]

Making this assumption enables the differentiation in tax rates to be carried through to the analysis, which allows more precise measurement of ECRs, particularly in terms of the percentage of emissions subject to no price signal.

The information on emissions from fuel use for 30 individual users from the TEU database can be grouped into the six sectors as described in Section A.3.1. To incorporate the estimates of ETS coverage for each sector into the TEU database, it is necessary to translate the estimates for each of the six to the individual users in the TEU database. For five of the six sectors – road transport, offroad transport, commercial and residential, agriculture and fishing, and electricity – the match is fairly precise, as the TEU database is not heavily disaggregated by individual users within these sectors and any ETS coverage is rare and where existing, low. However within the industrial sector, the TEU database provides information on the tax rates and coverage for 18 individual industrial users (also referred to as individual industries). These 18 individual industries have very different fuel and tax characteristics and can also differ strongly in the degree to which their emissions are subject to emissions trading systems.

Determining the ECR coverage for the industry sector is more complex than for the other five sectors. Precise matching of the individual industries in the TEU database to the national data on emissions subject to each ETS is not possible due to the difficulty of assigning auto-generation emissions to individual industries as described in Section A.2.2.2. However, applying the estimate of ETS coverage for the industry sector evenly across each of these individual industries is likely to be inaccurate, given their different fuel use characteristics and differing degrees of ETS coverage.

In the absence of more precise information on how the ETS applies individually to these industries, there is a range of uncertainty on how taxes interact with the applicable ETS. Conceptually, it is possible that the coverage of taxes and emissions trading systems entirely overlaps (i.e. all taxed emissions are subject to the ETS, or vice versa), or that their coverage is entirely separate (i.e. no taxed emissions are subject to the ETS). With complete overlap, the ECR coverage is at its possible minimum; with no overlap, the ECR coverage is at its possible maximum. The true ECR coverage must exist within the range of minimum and maximum possible coverage. In many cases, additional ETS-specific information can reduce this range of uncertainty leading to a more exact estimate of ECR coverage for the industry sector.

More precisely, the minimum share of emissions that are subject to a price signal (either a tax or an ETS) occurs when all taxed emissions are subject to an ETS or when all emissions subject to an ETS are taxed, i.e. when emissions subject to tax and ETS entirely overlap. In this case, the minimum ECR coverage equals the maximum of either the share of emissions that are taxed or that are subject to an ETS. Conversely, ECR coverage is at its maximum when all taxed emissions and emissions subject to an ETS do not overlap; that is, that no emissions are subject to both a tax and an ETS. In this case the maximum ECR coverage is the sum of taxed emissions and emissions subject to an ETS.

Figure A.2 illustrates the possible minimum and maximum coverage graphically. The area of each box represents the total CO_2 emissions of the industry sector in a country. The ECR coverage is minimal when taxed emissions and emissions subject to the ETS completely overlap as shown in the left panel of Figure A.2 and equals the maximum of the emissions that are taxed (as shown in the figure) or that are subject to an ETS (not shown in the figure). When taxed emissions and emissions subject to an ETS do not overlap at all, as shown in the right panel of Figure A.2, the ECR coverage is at its maximum and equals the sum of taxed emissions and emissions subject to an ETS.

Figure A.2. **Emissions subject to either a tax or an ETS**

Total overlap – minimum ECR coverage (LHS), total separation – maximum ECR coverage) (RHS)

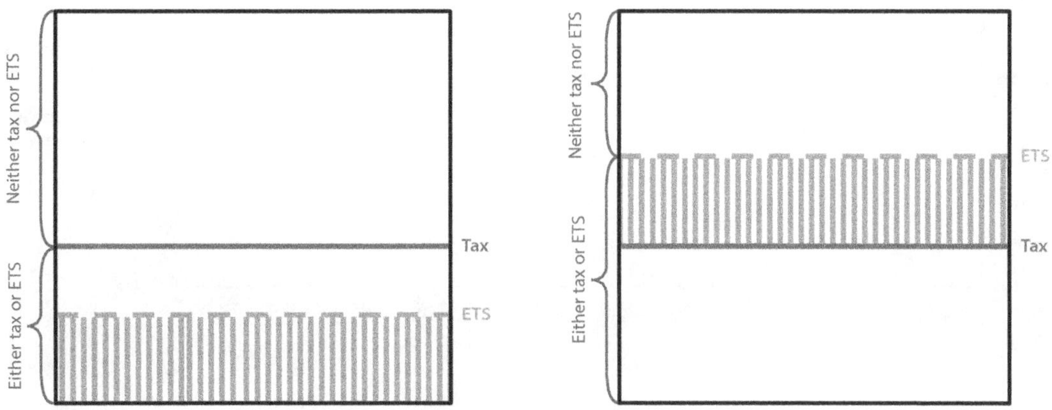

Note: The area of the box represents total emissions from the industry sector.

The range of uncertainty can be further narrowed by several constraints. Firstly, the ECR coverage can never be greater than one or less than zero. In systems where the coverage of one instrument is high, this in itself restricts the range of uncertainty considerably. Secondly, while it is not possible to precisely estimate the ETS coverage of each individual industry, additional information on the empirical coverage of some individual industries is available. For example, verified emissions data show that some industries are not subject to an ETS. As emissions from these individual industries are certainly not covered by an ETS, the uncertainty decreases. Thirdly, the level of uncertainty is further reduced by estimating a maximum level of possible ETS coverage (i.e. an overestimate of true coverage) for most of the other individual industries, which reduces the amount of emissions in the industry sector that can potentially be subject to both taxes and the ETS.

The approach to reduce the range of uncertainty described above is applied to countries that are part of the EU ETS, the Swiss ETS and the North-American emissions trading systems. The minimum and maximum coverage for other emissions trading systems are calculated without using empirical estimates of coverage of individual industries.

The point estimates of ECR coverage used in the analysis lie within this narrowed range of uncertainty. The estimates per fuel type are produced by reference to an assumption already described above that emissions from each fuel are assumed to be covered by an ETS to the same extent as the individual industry which uses that fuel. The tax coverage of each fuel will determine the degree of overlap or separation of tax and ETS coverage for that fuel. The point estimate of ECR coverage can then be determined by the respective fuel use of each individual industry.

The point estimates for the ECR coverage of each individual industry can be combined to generate the ECR coverage of the entire industry sector in each country. Figure A.3 shows the minimum and maximum ECR coverage of the industry sector by country, together with the point used in the estimates, as a percentage of total economy-wide CO_2 emissions from energy use.

Figure A.3. **Minimum and maximum industry ECR coverage and range used in analysis**

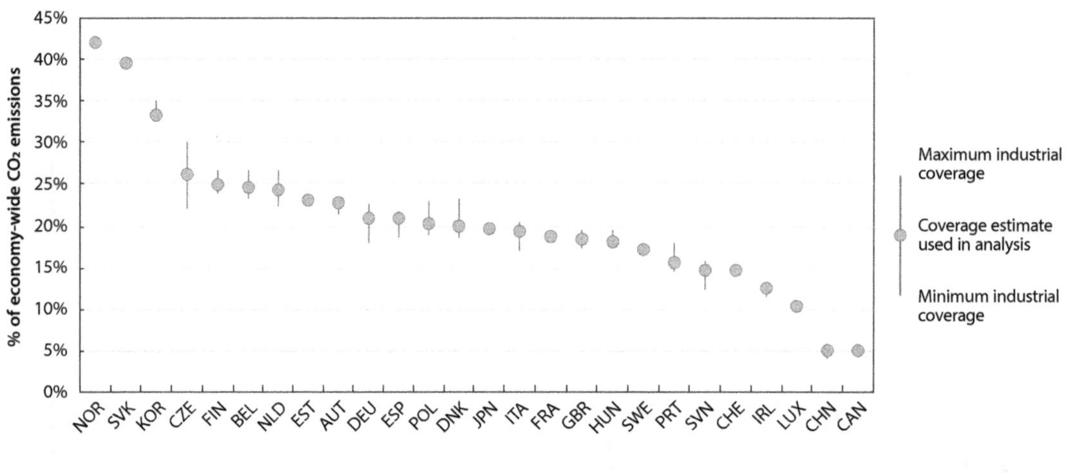

A.3.3. Adjusting carbon tax coverage in the TEU database

Another adjustment that must be made to the TEU database concerns carbon tax coverage. The original database included carbon tax rates as if any relevant ETS did not apply. In many countries with a carbon tax, however, the carbon tax is removed or reduced where the ETS applies. To take account of this, the rates of carbon taxes applied to the covered fuels are adjusted once the ETS coverage has been identified.

Table A.3 describes the treatment of carbon taxes in the presence of ETS coverage.

Table A.3. **Adjustments to carbon taxes made for covered fuels**

Jurisdiction	Treatment of carbon tax where ETS applies
Denmark	Fully removed
Finland	Not removed
Iceland	Not removed
Ireland	Fully removed
Norway	Fully removed, except for domestic aviation and energy transformation
Slovenia	Not removed
Sweden	Fully removed
Switzerland	Fully removed

A.3.4. Estimating effective carbon rates

Estimating ECRs requires several steps. First, CO_2 emissions from the TEU database, adjusted as described in Section A.3.1, are divided into two parts: a base that is covered by an ETS and a base that is not covered by an ETS. Second, the ECR is determined.

The base that is covered by an ETS is calculated as described in Section A.2 for all sectors except industry. For the industry sector, ETS coverage is estimated as described in Section A.3.2. The ECR for the base that is covered by an ETS is calculated as the sum of the applicable tax rate from the TEU database, using the adjusted rate of carbon taxes according to Section A.3.3, and the tradable emission permit price. The ECR for the base that is not covered by an ETS is calculated as the applicable tax rate as reported in the TEU database.

Tables A.4 and A.5 present for 41 OECD and G20 countries the ECR coverage for three carbon rate levels (EUR 0, EUR 5 and EUR 30 per tonne of CO_2) and average ECRs weighted by the amount of CO_2 emissions in each sector, together with weighted and simple means for all countries. Average ECRs are calculated by adding together taxes weighted with tax coverage and emission permit prices weighted with ETS coverage for the respective sector or economy. The tables also show the proportion of total CO_2 emissions in each country, ETS coverage and the average effective rates for each of the three ECR components (carbon taxes, other specific taxes on energy use, and tradable emission permit prices). Note that average rates are different from the actual rates that energy users have to pay.

All results are shown both inclusive and exclusive of CO_2 emissions from biomass. To give a clearer picture of the impact of each type of price mechanism, results for road transport are shown separately from the results for other sectors. Road is treated separately due to the high tax rates applied to road transport and the low coverage of emissions trading systems in this sector. Section B.2 in Annex B shows average ETS prices and ETS coverage for each of the six sectors in each country with an ETS.

Table A.4. **Composition of average effective carbon rates, including CO_2 emissions from biomass**

	All non-road energy									Road energy								
	Share of eco. base	ETS coverage	Average CO_2 tax	Average other tax	Average ETS rate	Average ECR	ECR coverage > EUR 0/ tCO_2	ECR coverage > EUR 5/ tCO_2	ECR coverage > EUR 30/ CO_2	Share of eco. base	ETS coverage	Average CO_2 tax	Average other tax	Average ETS rate	Average ECR	ECR coverage > EUR 0/ tCO_2	ECR coverage > EUR 5/ tCO_2	ECR coverage > EUR 30/ tCO_2
	% of emissions	% of emissions	EUR/tCO_2	EUR/tCO_2	EUR/tCO_2	EUR/tCO_2	% of emissions	% of emissions	% of emissions	% of emissions	% of emissions	EUR/tCO_2	EUR/tCO_2	EUR/tCO_2	EUR/tCO_2	% of emissions	% of emissions	% of emissions
AUS	82%	0%	0.00	2.44	0.00	2.44	6%	6%	3%	18%	0%	0.00	106.56	0.00	106.56	99%	99%	96%
AUT	75%	48%	0.00	17.98	2.22	20.20	57%	57%	26%	25%	0%	0.00	164.35	0.00	164.35	94%	94%	93%
BEL	79%	46%	0.00	4.67	2.74	7.41	75%	58%	4%	21%	0%	0.00	165.66	0.00	165.66	95%	95%	95%
CAN	73%	5%	1.42	1.56	0.40	3.38	23%	23%	1%	27%	22%	2.41	26.64	1.55	30.61	100%	100%	72%
CHL	84%	0%	0.00	-0.04	0.00	-0.04	0%	0%	0%	16%	0%	0.00	78.16	0.00	78.16	100%	100%	100%
CZE	84%	55%	0.00	2.95	3.77	6.72	74%	62%	1%	16%	0%	0.00	172.25	0.00	172.25	95%	95%	95%
DNK	77%	56%	3.97	40.83	2.60	47.40	59%	59%	39%	23%	0%	11.04	179.31	0.00	190.35	94%	94%	94%
EST	87%	65%	0.00	5.37	4.09	9.46	77%	65%	5%	13%	0%	0.00	160.43	0.00	160.43	99%	99%	99%
FIN	87%	69%	2.05	19.22	2.47	23.74	55%	54%	35%	13%	0%	44.24	171.87	0.00	216.11	100%	100%	100%
FRA	67%	41%	0.00	7.16	2.52	9.67	63%	62%	2%	33%	0%	0.00	179.87	0.00	179.87	100%	100%	100%
DEU	82%	63%	0.00	19.33	4.04	23.37	88%	85%	37%	18%	0%	0.00	219.48	0.00	219.48	100%	100%	100%
GRC	82%	77%	0.00	15.30	5.43	20.73	89%	89%	20%	18%	0%	0.00	241.36	0.00	241.36	100%	100%	100%
HUN	77%	50%	0.00	1.72	3.22	4.95	62%	57%	0%	23%	0%	0.00	137.39	0.00	137.39	96%	96%	96%
ISL	58%	2%	5.93	12.08	0.12	18.13	45%	45%	0%	42%	0%	13.63	151.94	0.00	165.57	99%	99%	99%
IRL	73%	60%	5.21	11.29	4.07	20.57	85%	85%	12%	27%	0%	19.70	190.88	0.00	210.58	100%	100%	100%
ISR	75%	60%	0.00	22.43	3.91	26.34	90%	89%	21%	25%	0%	0.00	239.39	0.00	239.39	100%	98%	98%
ITA	82%	0%	0.00	20.41	0.00	20.41	91%	21%	18%	18%	0%	0.00	242.69	0.00	242.69	100%	100%	100%
JPN	85%	1%	0.00	6.60	1.08	7.68	80%	53%	2%	15%	0%	0.00	188.29	0.00	188.29	100%	98%	98%
KOR	87%	90%	0.00	3.86	5.90	9.76	91%	91%	3%	13%	0%	0.00	153.25	0.00	153.25	99%	99%	96%
LUX	31%	31%	0.00	3.89	1.94	5.83	88%	62%	0%	69%	0%	0.00	135.50	0.00	135.50	98%	98%	98%
MEX	69%	0%	0.00	0.22	0.00	0.22	3%	3%	0%	31%	0%	0.00	8.14	0.00	8.14	99%	99%	0%
NLD	80%	56%	0.00	51.06	3.57	54.63	89%	88%	57%	20%	0%	0.00	224.82	0.00	224.82	100%	100%	100%
NZL	67%	49%	0.00	0.49	0.66	1.15	53%	1%	1%	33%	81%	0.00	89.09	1.08	90.17	100%	56%	56%
NOR	76%	71%	5.68	36.75	4.31	46.74	75%	75%	19%	24%	0%	34.82	204.53	0.00	239.34	100%	100%	99%
POL	85%	67%	0.00	6.15	4.46	10.61	71%	65%	3%	15%	0%	0.00	130.64	0.00	130.64	95%	95%	95%
PRT	72%	67%	0.00	2.81	3.77	6.58	68%	57%	1%	28%	0%	0.00	156.05	0.00	156.05	95%	95%	95%
SVK	83%	52%	0.00	13.14	3.13	16.27	77%	62%	22%	17%	0%	0.00	155.70	0.00	155.70	96%	96%	96%
SVN	62%	45%	2.63	10.87	3.06	16.56	68%	67%	16%	38%	0%	0.00	151.50	0.00	151.50	97%	97%	97%
ESP	74%	62%	0.00	6.98	4.14	11.12	74%	69%	8%	26%	0%	0.00	135.12	0.00	135.12	92%	92%	92%
SWE	78%	78%	3.09	26.08	1.49	30.66	36%	36%	18%	22%	0%	107.92	98.56	0.00	206.48	91%	91%	91%
CHE	64%	24%	12.99	4.68	3.80	21.47	73%	72%	42%	36%	0%	0.00	251.71	0.00	251.71	100%	100%	100%
TUR	85%	0%	0.00	7.58	0.00	7.58	62%	21%	5%	15%	0%	0.00	218.57	0.00	218.57	100%	100%	100%
GBR	77%	63%	0.13	9.89	4.26	14.28	78%	77%	8%	23%	0%	0.00	280.57	0.00	280.57	100%	100%	100%
USA	72%	7%	0.00	0.34	0.42	0.76	13%	7%	0%	28%	11%	0.00	17.25	1.13	18.37	100%	100%	0%
ARG	77%	0%	0.00	3.70	0.00	3.70	66%	25%	3%	23%	0%	0.00	130.92	0.00	130.92	99%	99%	99%
BRA	72%	0%	0.00	1.78	0.00	1.78	20%	18%	0%	28%	0%	0.00	9.10	0.00	9.10	81%	81%	0%
CHN	94%	10%	0.00	1.05	0.49	1.55	12%	8%	3%	6%	0%	0.00	41.98	0.00	41.98	95%	95%	95%
IND	93%	0%	0.00	0.96	0.00	0.96	50%	4%	0%	7%	0%	0.00	29.12	0.00	29.12	97%	97%	24%
IDN	83%	0%	0.00	0.00	0.00	0.00	0%	0%	0%	17%	0%	0.00	13.94	0.00	13.94	99%	99%	0%
RUS	92%	0%	0.00	0.00	0.00	0.00	5%	0%	0%	8%	0%	0.00	0.05	0.00	0.05	99%	0%	0%
ZAF	91%	0%	0.00	2.95	0.00	2.95	51%	2%	2%	9%	0%	0.00	122.08	0.00	122.08	100%	100%	100%
Weighted mean	85%	14%	0.09	3.05	0.90	4.17	30%	19%	4%	15%	3%	0.62	72.31	0.24	67.21	98%	94%	46%
Simple mean	77%	36%	1.05	9.92	2.15	13.12	57%	46%	11%	23%	3%	5.70	140.85	0.09	146.64	98%	94%	82%

StatLink http://dx.doi.org/10.1787/888933391050

Table A.5. Composition of average effective carbon rates, excluding CO_2 emissions from biomass

	All non-road energy									Road energy								
	Share of eco. base	ETS coverage	Average CO_2 tax	Average other tax	Average ETS rate	Average ECR	ECR coverage > EUR 0/ tCO_2	ECR coverage > EUR 5/ tCO_2	ECR coverage > EUR 30/ CO_2	Share of eco. base	ETS coverage	Average CO_2 tax	Average other tax	Average ETS rate	Average ECR	ECR coverage > EUR 0/ tCO_2	ECR coverage > EUR 5/ tCO_2	ECR coverage > EUR 30/ CO_2
	% of emissions	% of emissions	EUR/tCO_2	EUR/tCO_2	EUR/tCO_2	EUR/tCO_2	% of emissions	% of emissions	% of emissions	% of emissions	% of emissions	EUR/tCO_2	EUR/tCO_2	EUR/tCO_2	EUR/tCO_2	% of emissions	% of emissions	% of emissions
AUS	81%	0%	0.00	2.60	0.00	2.60	6%	6%	3%	19%	0%	0.00	107.70	0.00	107.70	100%	100%	97%
AUT	68%	47%	0.00	25.84	3.42	29.27	78%	78%	40%	32%	0%	0.00	175.48	0.00	175.48	100%	100%	100%
BEL	77%	43%	0.00	4.48	3.15	7.63	79%	59%	4%	23%	0%	0.00	172.97	0.00	172.97	99%	99%	99%
CAN	72%	6%	1.61	1.76	0.45	3.82	26%	26%	1%	28%	22%	2.50	26.05	1.61	30.16	100%	100%	71%
CHL	76%	0%	0.00	-0.07	0.00	-0.07	0%	0%	0%	24%	0%	0.00	78.16	0.00	78.16	100%	100%	100%
CZE	83%	59%	0.00	3.34	4.26	7.60	84%	71%	2%	17%	0%	0.00	181.23	0.00	181.23	100%	100%	100%
DNK	69%	59%	6.50	56.21	4.24	66.95	88%	88%	55%	31%	0%	11.74	190.66	0.00	202.40	100%	100%	100%
EST	84%	74%	0.00	6.77	5.36	12.13	97%	82%	7%	16%	0%	0.00	161.28	0.00	161.28	100%	100%	100%
FIN	78%	67%	4.05	29.87	4.87	38.79	90%	89%	50%	22%	0%	45.47	171.78	0.00	217.25	100%	100%	100%
FRA	63%	44%	0.00	8.60	3.18	11.78	75%	74%	3%	37%	0%	0.00	179.19	0.00	179.19	100%	100%	100%
DEU	81%	64%	0.00	20.77	4.63	25.40	94%	92%	42%	19%	0%	0.00	215.59	0.00	215.59	100%	100%	100%
GRC	81%	82%	0.00	16.66	5.93	22.59	96%	96%	22%	19%	0%	0.00	242.61	0.00	242.61	100%	100%	100%
HUN	74%	54%	0.00	2.00	3.88	5.88	70%	69%	0%	26%	0%	0.00	143.33	0.00	143.33	100%	100%	100%
ISL	59%	2%	5.93	12.08	0.12	18.13	45%	45%	0%	41%	0%	13.70	152.73	0.00	166.43	100%	100%	100%
IRL	72%	59%	5.51	11.93	4.30	21.75	90%	89%	13%	28%	0%	20.19	188.85	0.00	209.05	100%	100%	100%
ISR	74%	61%	0.00	24.35	4.42	28.77	96%	95%	23%	26%	0%	0.00	236.88	0.00	236.88	100%	98%	98%
ITA	82%	0%	0.00	20.46	0.00	20.46	91%	21%	18%	18%	0%	0.00	242.69	0.00	242.69	100%	100%	100%
JPN	85%	1%	0.00	6.72	1.12	7.84	80%	55%	2%	15%	0%	0.00	188.29	0.00	188.29	100%	98%	98%
KOR	87%	90%	0.00	3.92	5.98	9.90	93%	92%	3%	13%	0%	0.00	154.90	0.00	154.90	100%	100%	97%
LUX	30%	29%	0.00	4.19	2.09	6.28	95%	67%	0%	70%	0%	0.00	138.45	0.00	138.45	100%	100%	100%
MEX	66%	0%	0.00	0.25	0.00	0.25	4%	4%	0%	34%	0%	0.00	8.14	0.00	8.14	99%	99%	0%
NLD	79%	54%	0.00	50.12	3.92	54.04	90%	90%	55%	21%	0%	0.00	222.03	0.00	222.03	100%	100%	100%
NZL	62%	63%	0.00	0.62	0.84	1.46	68%	1%	1%	38%	81%	0.00	89.22	1.09	90.30	100%	56%	56%
NOR	72%	76%	7.26	38.27	5.51	51.03	95%	95%	23%	28%	0%	36.26	208.57	0.00	244.83	100%	100%	100%
POL	85%	70%	0.00	6.79	5.06	11.85	77%	73%	3%	15%	0%	0.00	137.74	0.00	137.74	100%	100%	100%
PRT	67%	72%	0.00	3.78	5.22	9.00	85%	78%	2%	33%	0%	0.00	164.45	0.00	164.45	100%	100%	100%
SVK	82%	49%	0.00	14.61	3.52	18.13	83%	69%	25%	18%	0%	0.00	162.81	0.00	162.81	100%	100%	100%
SVN	54%	60%	3.71	15.28	4.31	23.30	95%	94%	23%	46%	0%	0.00	155.66	0.00	155.66	100%	100%	100%
ESP	73%	65%	0.00	7.80	4.67	12.47	80%	78%	9%	27%	0%	0.00	146.82	0.00	146.82	100%	100%	100%
SWE	54%	67%	10.01	28.99	4.82	43.83	87%	87%	27%	46%	0%	117.69	107.48	0.00	225.17	99%	99%	99%
CHE	57%	20%	17.30	6.22	5.06	28.58	97%	96%	56%	43%	0%	0.00	252.17	0.00	252.17	100%	100%	100%
TUR	85%	0%	0.00	8.05	0.00	8.05	66%	23%	6%	15%	0%	0.00	218.72	0.00	218.72	100%	100%	100%
GBR	77%	62%	0.14	9.95	4.50	14.59	78%	77%	8%	23%	0%	0.00	279.97	0.00	279.97	100%	100%	100%
USA	72%	8%	0.00	0.36	0.45	0.81	13%	8%	0%	28%	12%	0.00	16.85	1.19	18.04	100%	100%	0%
ARG	77%	0%	0.00	3.93	0.00	3.93	68%	27%	3%	23%	0%	0.00	135.29	0.00	135.29	100%	100%	100%
BRA	61%	0%	0.00	2.73	0.00	2.73	30%	26%	0%	39%	0%	0.00	10.98	0.00	10.98	98%	98%	0%
CHN	93%	11%	0.00	1.18	0.55	1.73	14%	9%	3%	7%	0%	0.00	42.24	0.00	42.24	95%	95%	95%
IND	90%	0%	0.00	1.42	0.00	1.42	74%	5%	0%	10%	0%	0.00	29.21	0.00	29.21	98%	98%	24%
IDN	73%	0%	0.00	0.00	0.00	0.00	0%	0%	0%	27%	0%	0.00	14.13	0.00	14.13	100%	100%	0%
RUS	92%	0%	0.00	0.00	0.00	0.00	5%	0%	0%	8%	0%	0.00	0.05	0.00	0.05	99%	0%	0%
ZAF	90%	0%	0.00	3.44	0.00	3.44	59%	2%	2%	10%	0%	0.00	122.08	0.00	122.08	100%	100%	100%
Weighted mean	84%	15%	0.11	3.31	1.04	4.62	34%	21%	4%	16%	3%	0.42	73.1	0.26	68.73	99%	96%	47%
Simple mean	74%	37%	1.51	11.37	2.68	15.56	67%	55%	13%	26%	3%	6.04	143.25	0.09	149.39	100%	96%	84%

StatLink http://dx.doi.org/10.1787/888933391067

Notes

1. The 2013 and 2015 TEU reports show tax levels and coverage as at 1 April 2012 for all countries except Australia and Brazil (1 July 2012) and South Africa (4 April 2012). Data on energy use and emissions for 2009 were used, as they were the most recently available at the time when the initial publication was prepared.

   Information for Australia and Brazil in the TEU database initially showed tax rates at 1 July 2012. This was to better reflect the current tax system, because both tax systems had changed shortly after the 1 April 2012 date shown for other countries. However, since this date, both countries have reverted to tax settings more similar to the tax system in force at 1 April 2012: Australia, by removing the carbon price, and Brazil, by reinstating CIDE for transport fuels. Hence, 1 April 2012 is also used for these countries in the present report. In addition, a coding error was corrected for the Indonesian tax rates. No other changes to tax rates have been made.

2. Exceptions include Switzerland, where no auctions were held in 2013, and some of the Chinese emissions trading systems because 2014 is used consistently for all systems in China. Further, in California and Québec, the extension of ETS in 2015 to cover fossil fuel distributors has been included in the analysis.

3. Except for biomass, which is treated differently as described in Section A.2.2.4.

4. The same price has been used in the calculations for both Japanese systems.

5. This rate is halved because one permit is surrendered for every two tonnes of CO_2 emissions.

6. In the TEU database, data on energy use is taken from the *Extended World Energy Balances* (IEA, 2015). Energy use is converted into carbon emissions (in tonnes of CO_2), using the 2006 IPCC emission factors.

7. If emissions subject to an ETS are recorded at a facility level and if the corresponding facility is involved in different economic activities, emissions are attributed to the facility's main activity.

8. This is the case for all countries except China, where publicly available information does not allow the amount of auto-generation of electricity in total electricity generation to be identified at the level of the seven pilot systems. For China, therefore, emissions from the auto-generation of electricity have remained together with emissions from main electricity generation, in both the numerator and the denominator. In practical terms, auto-generation of electricity in China only plays a prominent role in aluminium production, which is mainly located outside the seven pilot systems. Hence, this deviation from the general approach likely does not affect results significantly in practice.

9. For California, data from the California Greenhouse Gas Emission Inventory provided by California's Environmental Protection Agency's Air Resources Board (2015) is used in addition to UNFCCC reports.

10. The Swiss ETS covers the same greenhouse gases as the EU ETS, but due to difference in the date used (2013 for the Swiss ETS and 2012 for the EU ETS), the treatment of greenhouse gases differs.

11. In New Zealand, emissions from liquid fuels and from coal include only CO_2 emissions.

12. In addition to the UNFCCC reports, for California data from the California Greenhouse Gas Emission Inventory provided by California's Environmental Protection Agency's Air Resources Board (2015) is used.

13. This change is due to the difficulty in separating industrial and auto-generation emissions in national data on verified emissions; see Section A.2.2.2 for further information.

14. This is not required for the ETS in New Zealand, where facility-based data is not used to assess coverage. Assumptions about the coverage of emissions from biomass also differ between systems and are described in Section A.2.2.4.

References

California Environmental Protection Agency Air Resources Board (2014), *2013 Compliance Obligation Detail for ARB's Cap-and-Trade Program* (database), www.arb.ca.gov/cc/capandtrade/2013compliancereport.xlsx (accessed 28 July 2015).

California Environmental Protection Agency Air Resources Board (2015), *California Greenhouse Gas Emission Inventory - 2015 Edition* (database), www.arb.ca.gov/cc/inventory/data/tables/ghg_inventory_by_ipcc_00-13_20150424.xlsx (accessed 28 July 2015).

European Commission (2014a), Classification of installations in the EUTL Registry based on NACE 4 statistical classification, http://ec.europa.eu/clima/policies/ets/cap/leakage/studies_en.htm (accessed 5 April 2015).

European Commission (2014b), Verified Emissions for 2013, http://ec.europa.eu/clima/policies/ets/registry/documentation_en.htm (accessed 5 April 2015).

Gouvernement du Québec (2013), "Quantité d'unités d'émission versée en allocation gratuite en 2013 et liste des émetteurs qui en ont bénéficié", *Gazette Officielle du Québec*, Vol. 148, No. 48, pp.1304-1305.

IEA (2015), Extended World Energy Balances, *IEA World Energy Statistics and Balances* (database), http://dx.doi.org/10.1787/data-00513-en.

Korean Ministry of Environment (2014), National Emissions Allocation Plan 2015-2017, www.gir.go.kr/home/board/read.do?pagerOffset=10&maxPageItems=10&maxIndexPages=10&searchKey=&searchValue=&menuId=19&boardId=49&boardMasterId=8&boardCategoryId (accessed 6 July 2015).

Li, J. et al. (2015), "Integrating Cap and Trade into Taxing Energy Use: Interim Report for China", report prepared for the OECD.

OECD (2015), *Taxing Energy Use 2015: OECD and Selected Partner Economies*, OECD Publishing, Paris, http://dx.doi.org/10.1787/9789264232334-en.

OECD (2013), *Taxing Energy Use: A Graphical Analysis*, OECD Publishing, Paris, http://dx.doi.org/10.1787/9789264183933-en.

RGGI (2015), Reports: Annual Emissions – Facility Level View, https://rggi-coats.org/eats/rggi/index.cfm?hc=ISkwICAK (accessed 27 August 2015).

Saitama Prefectural Government (2014), Saitama 2012 Emission Data, www.pref.saitama.lg.jp/a0502/sakugen.html (accessed 20 August 2015).

Swiss Emissions Trading Registry (2015), Public information, https://www.emissionsregistry.admin.ch/crweb/public/information.do (accessed 22 June 2015).

Swiss Federal Office for the Environment (2015), Email communication with the Swiss Federal Office for the Environment, 18 June 2015.

Tokyo Metropolitan Government (2012), "The Tokyo Cap-and-Trade Program: Results for first fiscal year of operation", Bureau of Environment, Tokyo Metropolitan Government, Tokyo, www.kankyo.metro.tokyo.jp/en/climate/attachement/Result%20of%20the%20First%20FY%20of%20the%20Tokyo%20CT%20Program_final.pdf (accessed 20 August 2015).

UNFCCC (2015), National Inventory Submission 2014, http://unfccc.int/national_reports/annex_i_ghg_inventories/national_inventories_submissions/items/8108.php (accessed 6 July 2015).

New Zealand Ministry for the Environment (2012), "ETS 2012 Facts and Figures", *Info 695*, New Zealand Ministry for the Environment, Wellington.

New Zealand Environmental Protection Authority (2012), "Climate Change Response Act 2012 report", report under the under section 89 and section 178A of the Climate Change Response Act 2002, New Zealand Environmental Protection Authority, www.epa.govt.nz/Publications/2012-13_Section_89_Report.pdf (accessed 27 August 2015).

Annex B

Description of emissions trading systems and results

Section B.1 describes the sixteen emissions trading systems for which coverage and permit prices were estimated: The Québec Cap-and-Trade System in Canada, the seven Chinese pilot systems (Beijing, Chongqing, Guangdong, Hubei, Shanghai, Shenzhen, Tianjin), the European Union Emissions Trading System, the systems in Saitama and Tokyo (Japan), Korea, New Zealand, Switzerland, and the California Cap-and-Trade Program as well as the Regional Greenhouse Gas Initiative in the United States. Section B.2 shows results on average effective ETS prices and ETS coverage by sector. Section B.3 lists the ETS-specific adjustment and assumptions that were made in the calculations.

B.1. Description of emissions trading systems used in the analysis

Table B.1 summarises the main features of the sixteen emissions trading systems included in this report.

Table B.1. **Description of emissions trading systems used in the analysis**

Country or region	Jurisdictions covered	Date trading started	Year used in the estimation	Greenhouse gas emissions covered	Sectors and activities covered	Threshold for inclusion of emitting facility (per year)	Number of facilities	Price ceiling or floor	Treatment of biomass
CAN	Quebec	Jan 2013	2013	CO_2, CH_4, N_2O, SF_6, HFCs, PFCs, NF_3	• Mining and quarrying, • Oil & natural gas extraction, • Electricity generation, • Manufacturing, • Fossil fuel distribution (2015)	25 000 tCO_2-eq	76	• Price floor: CAD 10.75 (2013), increasing annually by 5% plus rate of inflation, • Price ceiling: release of allowances from reserve account at CAD 40, 45 and 50	Not covered
CHN	Beijing	Nov 2013	2014 (partial)	CO_2	• Electricity & heat generation, • Industries, • Services	10 000 tCO_2	490		
	Chongqing	June 2014	2014 (partial)	CO_2, CH4, N2O, HFCs, PFCs, SF6	• Electricity & heat generation, • Industries	20 000 tCO_2-eq	242		
	Guangdong	Dec 2013	2014 (partial)	CO_2	• Electricity & heat generation, • Industries	20 000 tCO_2	202		
	Hubei	April 2014	2014 (partial)	CO_2	• Electricity & heat generation, • Industries	15 000 tCO_2	138		
	Shanghai	Nov 2013	2014 (partial)	CO_2	• Electricity & heat generation, • Industries, • Services	20 000 tCO_2	191		
	Shenzhen	June 2013	2014 (partial)	CO_2	• Electricity & heat generation, • Water supply, • Manufacturing, • Buildings	5 000 tCO_2	635 (+197 public buildings)		
	Tianjin	Dec 2013	2014 (partial)	CO_2	• Power & heat generation, • Industries	20 000 tCO_2	114		
EU	23 countries	Jan 2005	2012	CO_2	• Energy activities, • Ferrous metal, • Cement, glass and ceramics, • Pulp and paper • Aviation	Depends on activity, e.g. 20 MW thermal input for combustion activities, see European Commission (2003, Annex I) for details	> 11000		No need to surrender permits
JPN	Tokyo	April 2010	2012	CO_2	Large commercial and industrial facilities	1 500 kl energy in crude oil equivalents	1232		No need to surrender permits
	Saitama	April 2011	2012	CO_2	Large commercial and industrial facilities	1 500 kl energy in crude oil equivalents	581		No need to surrender permits

Table B.1. **Description of emissions trading systems used in the analysis** *(continued)*

Country or region	Jurisdictions covered	Date trading started	Year used in the estimation	Greenhouse gas emissions covered	Sectors and activities covered	Threshold for inclusion of emitting facility (per year)	Number of facilities	Price ceiling or floor	Treatment of biomass
KOR	National	Jan 2015	2015 (partial)	CO_2, CH_4, N_2O, HFCs, PFCs, SF_6	• Electricity generation, • Manufacturing, • Water supply, • Waste & sewerage, • Construction, • Air transport (5 airlines)	125 000 tCO_2-eq (companies) 25 000 tCO_2-eq (installations)	525	President can enact market stabilisation measures (price ceiling and floor, additional release of permits)	No need to surrender permits
NZL	National	Jan 2008 (extended to more sectors in following years)	2012	CO_2, CH_4, N_2O, HFCs, PFCs, SF_6	• Liquid fuel producers, • Electricity & heat generation, • Industrial processes, • Waste, • Synthetic GHG	Upstream coverage for each sector	2483	• Mandatory participants can surrender one permit for every two tonnes of emissions. • Price ceiling: NZD 12.5	Not covered
CHE	National	Jan 2008	2014	CO_2 (since 2013: N_2O, PFKW)	• Cement, • Chemicals & pharmaceuticals, • Paper production, • Metal production • District heating	20 MW (installed total rated thermal input)	55		No need to surrender permits
USA	California	Jan 2013	2013	CO_2, CH_4, N_2O, SF_6, HFCs, PFCs, NF_3	• Main industrial facilities, • Electricity generation & imports, • CO_2 suppliers, • Transport fuels (2015)	25 000 tCO_2-eq	387	• Price floor: USD 10 in 2012, increasing annually by 5% plus rate of inflation, • Price ceiling: release of allowances from reserve account at USD 40, 45 and 50	Not covered
	RGGI	Jan 2009	2012	CO_2	Electricity generation	25 MW (installed capacity)	170		Not covered if certain eligibility criteria apply, see RGGI (2008) for details

B.2. Permit prices and ETS coverage

Table B.2 shows the shares of ETS coverage by sector together with the average effective ETS prices for all countries and regions in which an ETS applies. For subnational emissions trading systems, the emissions covered by the ETS are given as national shares. The last column lists the price of tradable emission permits in 2012 EUR per tonne of CO_2.

Table B.2. **Permit prices and ETS coverage**

			ETS coverage						
			Road	Offroad	Industrial production	Agriculture & fishing	Residential & Commercial	Electricity	ETS Price (in EUR per tonne of CO_2)
AUT		Emissions covered by ETS (%)	0%	15%	39%	0%	0%	68%	7.24
		Average effective ETS price (price*coverage) in EUR per t CO_2	0.00	1.07	2.80	0.00	0.01	4.89	
BEL		Emissions covered by ETS (%)	0%	2%	50%	0%	0%	66%	7.24
		Average effective ETS price (price*coverage) in EUR per t CO_2	0.00	0.12	3.61	0.00	0.01	4.80	
CAN	Quebec	Emissions covered by ETS (%)	19%	9%	5%	1%	8%	1%	8.05
		Average effective ETS price (price*coverage) in EUR per t CO_2	1.55	0.70	0.39	0.09	0.68	0.06	
CHE		Emissions covered by ETS (%)	0%	7%	42%	0%	0%	0%	24.97
		Average effective ETS price (price*coverage) in EUR per t CO_2	0	1.78	10.61	0	0	0	
CHN	Beijing	Emissions covered by ETS (%)	0.0%	0.0%	0.5%	0.0%	0.7%	0.6%	6.37
		Average effective ETS price (price*coverage) in EUR per t CO_2	0.00	0.00	0.03	0.00	0.05	0.04	
	Chongqing	Emissions covered by ETS (%)	0.0%	0.0%	1.5%	0.0%	0.0%	0.9%	3.56
		Average effective ETS price (price*coverage) in EUR per t CO_2	0.00	0.00	0.06	0.00	0.00	0.03	
	Guangdong	Emissions covered by ETS (%)	0.0%	0.0%	2.7%	0.0%	0.0%	5.6%	7.20
		Average effective ETS price (price*coverage) in EUR per t CO_2	0.00	0.00	0.20	0.00	0.00	0. 40	
	Hubei	Emissions covered by ETS (%)	0.0%	0.0%	2.0%	0.0%	0.0%	1.9%	2.75
		Average effective ETS price (price*coverage) in EUR per t CO_2	0.00	0.00	0.06	0.00	0.00	0.6	
	Shanghai	Emissions covered by ETS (%)	0.0%	0.0%	1.4%	0.0%	3.7%	2.4%	4.19
		Average effective ETS price (price*coverage) in EUR per t CO_2	0.00	0.00	0.06	0.00	0.16	0.10	
	Shenzhen	Emissions covered by ETS (%)	0.0%	0.0%	0.01%	0.0%	0.0%	0.6%	8.52
		Average effective ETS price (price*coverage) in EUR per t CO_2	0.00	0.00	0.01	0.00	0.00	0.05	
	Tianjin	Emissions covered by ETS (%)	0.0%	0.0%	2.1%	0.0%	0.0%	1.2%	3.58
		Average effective ETS price (price*coverage) in EUR per t CO_2	0.00	0.00	0.07	0.00	0.00	0.04	
CZE		Emissions covered by ETS (%)	0%	15%	39%	0%	0%	97%	7.24
		Average effective ETS price (price*coverage) in EUR per t CO_2	0.00	1.06	2.83	0.00	0.02	7.01	
DNK		Emissions covered by ETS (%)	0%	10%	40%	2%	0%	77%	7.24
		Average effective ETS price (price*coverage) in EUR per t CO_2	0.00	0.71	2.89	0.16	0.01	5.57	
EST		Emissions covered by ETS (%)	0%	2%	20%	0%	0%	95%	7.24
		Average effective ETS price (price*coverage) in EUR per t CO_2	0.00	0.15	1.42	0.00	0.00	6.85	
FIN		Emissions covered by ETS (%)	0%	24%	35%	1%	0%	66%	7.24
		Average effective ETS price (price*coverage) in EUR per t CO_2	0.00	1.74	2.52	0.05	0.00	4.77	

Table B.2. **Permit prices and ETS coverage** *(continued)*

			ETS coverage						ETS Price (in EUR per tonne of CO_2)
			Road	Offroad	Industrial production	Agriculture & fishing	Residential & Commercial	Electricity	
FRA		Emissions covered by ETS (%)	0%	60%	59%	0%	1%	96%	7.24
		Average effective ETS price (price*coverage) in EUR per t CO_2	0.00	4.38	4.29	0.00	0.05	6.97	
DEU		Emissions covered by ETS (%)	0%	39%	58%	0%	0%	89%	7.24
		Average effective ETS price (price*coverage) in EUR per t CO_2	0.00	2.79	4.22	0.00	0.01	6.47	
GRC		Emissions covered by ETS (%)	0%	23%	92%	10%	0%	100%	7.24
		Average effective ETS price (price*coverage) in EUR per t CO_2	0.00	1.64	6.67	0.75	0.00	7.21	
HUN		Emissions covered by ETS (%)	0%	81%	71%	0%	0%	85%	7.24
		Average effective ETS price (price*coverage) in EUR per t CO_2	0.00	5.89	5.17	0.00	0.01	6.13	
ISL		Emissions covered by ETS (%)	0%	58%	0%	0%	0%	0%	7.24
		Average effective ETS price (price*coverage) in EUR per t CO_2	0.00	4.17	0.00	0.00	0.00	0.00	
IRL		Emissions covered by ETS (%)	0%	16%	60%	0%	0%	96%	7.24
		Average effective ETS price (price*coverage) in EUR per t CO_2	0.00	1.12	4.33	0.00	0.03	6.98	
ITA		Emissions covered by ETS (%)	0%	46%	66%	0%	0%	87%	7.24
		Average effective ETS price (price*coverage) in EUR per t CO_2	0.00	3.32	4.76	0.00	0.03	6.33	
JPN	Tokyo	Emissions covered by ETS (%)	0%	0%	0.2%	0%	0.3%	1.4%	85.99
		Average effective ETS price (price*coverage) in EUR per t CO_2	0.00	0.00	0.17	0.00	0.27	1.26	
	Saitama	Emissions covered by ETS (%)	0%	0%	0.6%	0.1%	0%	0.5%	85.99
		Average effective ETS price (price*coverage) in EUR per t CO_2	0.00	0.00	0.52	0.10	0.00	0.45	
KOR		Emissions covered by ETS (%)	0%	32%	91%	0%	87%	90%	6.65
		Average effective ETS price (price*coverage) in EUR per t CO_2	0.00	2.11	6.02	0.00	5.78	6.00	
LUX		Emissions covered by ETS (%)	0%	0%	61%	0%	0%	85%	7.24
		Average effective ETS price (price*coverage) in EUR per t CO_2	0.00	0.00	4.45	0.00	0.00	6.14	
NLD		Emissions covered by ETS (%)	0%	16%	53%	17%	1%	91%	7.24
		Average effective ETS price (price*coverage) in EUR per t CO_2	0.00	1.19	3.82	1.21	0.06	6.59	
NOR		Emissions covered by ETS (%)	0%	30%	83%	0%	0%	58%	7.24
		Average effective ETS price (price*coverage) in EUR per t CO_2	0.00	2.16	6.04	0.00	0.00	4.22	
NZL		Emissions covered by ETS (%)	81%	81 %	41%	71%	53%	54%	1.33
		Average effective ETS price (price*coverage) in EUR per t CO_2	1.08	1.09	0.55	0.95	0.71	0.72	
POL		Emissions covered by ETS (%)	0%	56%	67%	1%	0%	93%	7.24
		Average effective ETS price (price*coverage) in EUR per t CO_2	0.00	4.05	4.82	0.05	0.01	6.77	
PRT		Emissions covered by ETS (%)	0%	46%	42%	1%	0%	93%	7.24
		Average effective ETS price (price*coverage) in EUR per t CO_2	0.00	3.32	3.02	0.06	0.00	6.73	
SVK		Emissions covered by ETS (%)	0%	100%	46%	0%	0%	89%	7.24
		Average effective ETS price (price*coverage) in EUR per t CO_2	0.00	7.24	3.34	0.00	0.02	6.46	
SVN		Emissions covered by ETS (%)	0%	7%	51%	0%	0%	98%	7.24
		Average effective ETS price (price*coverage) in EUR per t CO_2	0.00	0.53	3.66	0.00	0.02	7.10	

Table B.2. **Permit prices and ETS coverage** *(continued)*

			ETS coverage						ETS Price (in EUR per tonne of CO_2)
			Road	Offroad	Industrial production	Agriculture & fishing	Residential & Commercial	Electricity	
ESP		Emissions covered by ETS (%)	0%	45%	53%	0%	1%	99%	7.24
		Average effective ETS price (price*coverage) in EUR per t CO_2	0.00	3.24	3.81	0.00	0.04	7.17	
SWE		Emissions covered by ETS (%)	0%	68%	23%	0%	0%	24%	7.24
		Average effective ETS price (price*coverage) in EUR per t CO_2	0.00	4.89	1.65	0.00	0.00	1.77	
GBR		Emissions covered by ETS (%)	0%	40%	56%	10%	1%	96%	7.24
		Average effective ETS price (price*coverage) in EUR per t CO_2	0.00	2.92	4.03	0.71	0.08	6.95	
USA	California	Emissions covered by ETS (%)	11%	2%	5%	8%	8%	4%	9.84
		Average effective ETS price (price*coverage) in EUR per t CO_2	1.13	0.24	0.49	0.77	0.79	0.38	
	RGGI	Emissions covered by ETS (%)	0%	0%	0.3%	0%	0%	4.5%	1.48
		Average effective ETS price (price*coverage) in EUR per t CO_2	0	0	0.03	0	0	0.24	

B.3. ETS-specific adjustments and assumptions

Results presented in Table B.2 are based on the general estimation methodology as detailed in Sections A.2 and A.3. In addition, the following specific adjustments and assumptions are made for the different emissions trading systems.

B.3.1. Quebec

Data on emissions subject to the Quebec Cap-and-Trade System in 2013 is from Quebec's Ministry of Sustainable Development, Environment and the Fight against Climate Change (Gouvernement du Quebec, 2013). The dataset reports the verified greenhouse gas emissions in 2013 of all facilities subject to the system._In addition, the following ETS-specific adjustments and assumptions are made:

- *Matching* of facilities to the six sectors discussed in the main body of the report is based on the North American Industry Classification System (NAICS). NAICS codes are self-reported in the dataset.
- *Non-CO_2 and industrial process emissions* are excluded based on Quebec information within Canada's 2013 UNFCCC emission inventory (Environment Canada, 2015).
- As of 2015, fossil fuel distributors are included in Quebec's Cap-and-Trade System, which has a significant effect on the *coverage* of the road and offroad transport sectors as well as that of the residential and commercial, and agriculture and fisheries sectors. The extension of coverage to these sectors is included in the estimate.

  To approximate the extension of coverage, Quebec's share in Canada's CO_2 emissions from energy use is calculated for each of the four additional sectors using 2013 data from Environment Canada (2015) and added to the estimates. This implies that the extension of coverage in 2015 is approximated using 2013 data and translated one-to-one into coverage in 2012, to be consistent with the TEU base data.

The methodology effectively assumes a stable relationship over time between the size of CO_2 emissions in the respective sectors in Quebec and Canada. It further assumes that supply and demand could have resulted in the same permit price, had the coverage been extended in 2012 already. Finally, the approximation yields an upper bound of the coverage as it assumes a cap-and-trade coverage of 100% in these four sectors.

- The average auction *price* in 2013 was CAD 10.75 per tonne of CO_2 (Gouvernement du Québec, 2014) which translates into EUR 8.05 when adjusted to the price level of 2012 using data from OECD (2010) and exchange rates from the TEU database.

B.3.2. People's Republic of China

Due to data limitations, the estimation of Chinese ETS coverage differs to some extent from the general methodology outlined in the report. The estimation follows Li et al. (2015) and proceeds in four major steps, and each may include several minor steps.

In the first major step, total emissions by sector are calculated for each municipality or region. Only sectors that are subject to an ETS are considered, i.e. electricity and heat generation, individual industries and services. Data on energy use by fuel and sector is from the 2013 Energy Balance Sheets of the *China Statistical Yearbook* (National Bureau of Statistics of China, 2014).

Because the data for the industry sector is reported at a very aggregate level, total emissions are split up into individual industries using shares from different sources:

- Structural information on energy consumption at a 42-sector level from 2012 is reported in provincial statistical yearbooks and used for Beijing, Guangdong, Hubei and Shanghai.
- Detailed energy use data per individual industry in 2013 from the *China Statistical Yearbook* is used for Tianjin.
- Energy consumption for main secondary industries in 2013 from the *China Statistical Yearbook* is used for Shenzhen.
- Energy consumption of large companies in 2013, which is mapped into individual industries, is used for Chongqing.

In the second major step total energy use is transformed into CO_2 emissions. The transformation into CO_2 emissions uses IPCC emission factors for fuels and fuel combustion rates from Ou et al. (2010) and is done for direct emissions and indirect emissions from electricity and from heat. Direct emissions are emissions generated during the combustion of fossil fuels in a sector. Indirect emissions from electricity are emissions from the production of electricity that is used by industries or services, but produced outside the municipality or region. Indirect emissions from heat are emissions generated during the production of heat that is used by industries and services but produced outside of the facility, be it inside or outside the municipality or region.

In the third major step the emissions that are subject to an ETS in China are estimated for all municipalities and regions. Depending on data availability, two approaches apply. First, for Chongqing, Guangdong, Hubei, Shanghai and Shenzhen publicly available information from subnational energy balances is matched with expert assessments to estimate ETS coverage. Second for Beijing and Tianjin ETS coverage is derived from publicly available information on energy use of companies. Emissions subject to the respective emissions trading systems are assessed separately for electricity and heat generation, individual industries as well as services. All indirect emissions from electricity that are subject to the ETS are counted towards the electricity sector. Similarly, indirect emissions from heat that are subject to the ETS are counted towards the industrial sector, which includes emissions from heat generation as described in Section A.3.1.

In the fourth major step the emissions subject to an ETS in the seven municipalities and regions are aggregated at the national level to represent China's covered emission in the six sectors outlined in Section A.1.3.

The general estimation methodology is applied as detailed in Sections A.2 and A.3. In addition, the following specific adjustments and assumptions are made:

- *Industrial process emissions* from lime and cement production are excluded from the estimation of the emissions trading systems in Beijing, Chongqing, Guangdong, Shanghai and Shenzhen using a uniform process emission factor (0.57 tonnes CO_2 per tonne of cement) from the public literature. Cement and lime production account for 87% of China's total industrial process emissions in 2005 (UNFCCC, 2015).
- *Auto-generation of electricity* in China mainly occurs in the aluminium sector (Li, 2015). Because the vast majority of Chinese aluminium is produced outside the area of the seven pilot systems, no adjustment for auto-electricity is made.
- To calculate *indirect CO_2 emissions from electricity*, an average value for the national carbon intensity of electricity production is used. This implicitly assumes that electricity consumed in different sectors and regions is produced based on the same energy mix.
- In *Shanghai*, emissions from air transport subject to the ETS are currently allocated to the commercial and residential sector.
- In *Shenzhen*, 197 public buildings are covered by the Shenzhen ETS, but no information is available to estimate the total sector coverage.
- *Prices:*

  Beijing's average price in 2014 amounts to CNY 52.64 (EUR 6.37) per tonne CO_2, when adjusted to the price level of 2012.

  Chongqing's average price in 2014 amounts to CNY 29.38 (EUR 3.56) per tonne CO_2, when adjusted to the price level of 2012.

  Guangdong's average price in 2014 amounts to CNY 59.43 (EUR 7.20) per tonne CO_2, when adjusted to the price level of 2012.

  Hubei's average price in 2014 amounts to CNY 22.68 (EUR 2.75) per tonne CO_2, when adjusted to the price level of 2012.

  Shanghai's average price in 2014 amounts to CNY 34.64 (EUR 4.19) per tonne CO_2, when adjusted to the price level of 2012.

  Shenzhen's average price in 2014 amounts to CNY 70.34 (EUR 8.52) per tonne CO_2, when adjusted to the price level of 2012.

Tianjin's average price in 2014 amounts to CNY 29.57 (EUR 3.58) per tonne CO_2, when adjusted to the price level of 2012.

- To determine the price signal sent by ETS in China an average price from all pilot systems has been calculated for each sector, weighted by the proportion of covered emissions from each system in that sector. Price information is based on market prices from the emission exchanges in the period January to August 2014 and adjusted for inflation (Federal Reserve, 2015).

  Expressed in 2012 prices, the weighted effective average price in China amounts to CNY 38.66 (EUR 4.68) in the industry sector, CNY 37.68 (i.e. EUR 4.56) in the commercial and residential sector, CNY 45.08 (i.e. EUR 5.46) in the electricity sector.[1]

B.3.3. EU

Data on verified emissions from more than 11 000 installations and their corresponding economic activity classification is from the European Commission (2014a, 2014b). In addition, the following specific assumptions are made:

- The Statistical Classification of Economic Activities in the European Community (NACE) classifies economic activities into 615 economic classes. Where available, NACE codes are used to *match* emissions from installations to effective carbon rate (ECR) sectors.[2] If no NACE code is available installations are matched to ECR sectors codes using a compulsory activity code, that derives from the list of activities included in the EU ETS shown in Table B.1. For electricity generation and heat generation installations, there is only one common activity code. To distinguish between emissions from electricity and heat generation from installations with no NACE code but a common activity code two steps are undertaken. First, a ratio of emissions from electricity generation installations to heat generation installations is built based on information from installations for which a NACE code is available. Second emissions from electricity and heat generation from installations with no NACE code but a common activity code, are attributed to electricity and heat generation according to the ratio calculated in the first step.
- The *electricity sector* is covered close to 100% in most countries, given that even small fossil fuel power plants (>= 20 MW capacity) have to take part in the ETS. For some countries coverage was initially estimated to exceed 100% in the electricity sector. This is a consequence of initially allocating all emissions from electricity generation under the ETS to the electricity sector, even though some of these emissions result from auto-electricity generation in the industry sector. NACE and activity codes only specify the main activity of an installation. A power plant owned by an industrial company that produces electricity for its own production, but at the same time also for the grid maybe coded simply as a power plant. All its emissions would then be coded as emissions from the electricity sector, even though some are from auto-electricity generation in the industry sector. Knowing that the coverage of the electricity sector cannot be above 100% and that the rules on inclusion of power plants onto the ETS imply close to 100% coverage of the electricity sector any emissions in excess of 100% coverage of the electricity sector are called overflow emissions and separated from the electricity sector. These overflow emissions are then allocated to the industry sector, given that they result from auto-generation of electricity in the industry sector.

- With respect to *aviation*, commercial flights within the EEA area are covered by the EU ETS as described above. The *Taxing Energy Use* database only takes emission from domestic flights into account. For that reasons it is assumed that all regularly scheduled domestic flights are covered by the EU ETS, which are 97% of all flights within the EEA according to a study for the European Aviation Safety Agency (EASA 2009).
- The average allowance *price* of all auctions in 2012 was EUR 7.24 per ton of CO_2 (European Energy Exchange, 2015)

B.3.4. Japan

B.3.4.1. Tokyo

Emissions by sector from are from the Tokyo Metropolitan Government (TMG, 2012c). In addition, the following system-specific assumptions are made:

- The share of *indirect emissions from electricity use*, i.e. emission from electricity bought from an electricity provider, in total emissions is calculated for each sector using Tokyo's greenhouse gas inventory (TMG, 2015). Thereby it is assumed that the share of indirect emissions from electricity use in total emissions is the same for covered and non-covered facilities in a given sector.
- 2012 emissions under the cap are currently not available by economic subsector. For this reason subsector data from the latest year available, i.e. 2010, are *scaled* to 2012 by the difference of aggregate emissions under the cap from 2010 to 2012. TMG confirmed by email that emission reductions between 2010 and 2012 are fairly similar for the industrial and the commercial sector.
- The detailed subsector data for 2010 has only been published as preliminary data. Meanwhile official data show higher aggregate emissions for 2010 than the preliminary data does. The preliminary by subsector data is therefore *scaled* to match the official aggregate data for 2010.
- In 2012 the TMG assessed the average *price* of an excess reduction credit to be between JPN 9 300 and JPN 9 600. One excess reduction credit allows emitting one tonne CO_2 (TMG, 2012b). For the calculation of effective carbon rates a central estimate of JPN 9 450 per tonne CO_2 was assumed.

B.3.4.2. Saitama

Emissions by sector from are from the Saitama Prefectural Government (SPG, 2014a). In addition, the following system-specific assumptions are made:

- The share of *indirect emissions from electricity use*, i.e. emission from electricity bought from an electricity provider, in total emissions is calculated for each sector using Saitama's greenhouse gas inventory (SPG, 2014b). Thereby it is assumed that the share of indirect emissions from electricity use in total emissions is the same for covered and non-covered facilities for a given sector.
- So far no *price* information for Saitama ETS credits is available. Given that the Saitama ETS is linked with the Tokyo ETS via SME-credits it is assumed that the credit price from Tokyo also applies to Saitama. In 2012 the TMG assessed the average price of an excess reduction credit to be between JPN 9 300 and JPN 9 600.

One excess reduction credit allows emitting one tonne CO_2 (TMG, 2012b). For the calculation of effective carbon rates a central estimate of JPN 9 450 per tonne CO_2 was assumed.

B.3.5. Korea

Due to the recent introduction of the system, data of verified emissions by facility or company are not yet available. The 2015 allocation of emission permits to industrial sectors published by the Korean Ministry of Environment (2014) has instead been used to proxy for emissions by company sector, as described in Section A.2. Choosing allocated emissions by sector has two important implications for the Korean ECR estimate.

First, the absence of trade on the Korean emissions market could indicate that allocated permits will exceed verified emissions. Constructing shares of ETS coverage using allocated permits will thus likely overestimate ETS coverage, since the ETS permits allocated might be higher than verified emissions.

Second, as described in Section A.2, shares of ETS coverage are estimated using 2012 energy use data from the *IEA Extended World Energy Balances*. Constructing a share of the 2015 allocation of emissions over the 2012 emissions from energy use would necessarily result in an overestimate of ETS coverage, since the comparison does not account for emissions growth between 2012 and 2015. Consequently, the final allocation of emissions in 2015 to sectors was adjusted for emissions growth between 2012 and 2015.

Data on Korean emissions for the years between 2012 and 2015 could not be identified. Therefore, a simple average growth rate of the share of emissions between 2011 and 2012 (2.50%, yearly) was taken from the IEA Dataset of CO_2 Emissions from Fuel Combustion to scale back the 2015 allocation of emissions to sectors in 2012.

The data on allocation by industry sector is from the Korean Ministry of Environment (2014). The Ministry defines industry sectors by NACE industry codes in the same document, and sectors are matched to ECR sectors accordingly. In addition, the following ETS-specific assumptions and adjustments are made:

- *Industrial process emissions* are retrieved from the UNFCCC as submitted by the Korean authorities. As a non-Annex I country, Korea has different reporting requirements under the UNFCCC, limiting the precision of the data for individual industry sectors.
- A share of *non-GHG emissions* is calculated from UNFCCC data for all ECR sectors and deducted from the total permits allocated to sectors.
- It is assumed that *biomass* is covered by the scheme, but that its emissions factor is equal to zero.
- ETS spot *prices* are retrieved from the Korean emissions exchange (Korean Exchange KRX, 2015). The ETS price is calculated as the weighted average price between 12 January (1st trading day) and 1 July 2015, giving an average price of KRW 10 177. Adjusted for inflation and using *Taxing Energy Use* exchange rates for 2012, this yields EUR 6.65.
- The average permit *price* was KRW 10 177 per tonne of CO_2 in the first half of 2015 which translates into EUR 6.65 when adjusted to the price level of 2012 using data from OECD (2010) and exchange rates from the TEU database.

B.3.6. New Zealand

The coverage of the New Zealand ETS is estimated differently than for the other ETS. Given that the New Zealand ETS is an upstream ETS, applying to fuels (whether imported or produced) rather than to users, coverage is estimated by calculating which proportion of each fuel type is subject to the ETS. The proportion of ETS coverage for emissions from each fuel is assumed to apply uniformly to all users of that fuel. To calculate coverage at the user level total covered emissions from fuels used by each user are then divided by the total emissions for that user. Data on total covered emissions by user for the 2012 calendar year are published by the New Zealand Emission Unit Register (2015). In addition, the following assumptions are made:

- CO_2 *emissions* from energy use occur *primarily in the liquid fossil fuel and stationary energy sector.* Consequently, emissions from other sectors (e.g. industrial or agricultural) have been excluded from the estimation.
- The data retrieved from the New Zealand Emission Unit register list *emissions from industrial processes* separately from emissions from energy use, so they have directly been excluded from the estimation.
- It is assumed that *gas and liquid fuels* for export are *not covered* by the ETS, whereas *coal* for export is *covered.*
- Half of the average *price* of Emission Reduction Units (ERUs) from Joint Implementation projects in 2012 is used, to reflect the fact that 79% of surrendered permits in 2012 were ERUs (New Zealand Ministry for the Environment, 2012). The rate is halved because one permit is surrendered for every two tonnes of CO_2 emissions.
- The average permit *price* has been calculated as half the average ERU price for 2012: EUR 1.33 per tonne CO_2, or NZD 2.18 per tonne CO_2.

B.3.7. Switzerland

A list of ETS installations, as well as surrendered emissions by installation and year can be retrieved from the website of the Swiss Emissions Trading Registry (2015). A match of installations to sectors has been provided by the Swiss Federal Office for the Environment (Swiss Federal Office for the Environment 2015b). Where ambiguous, the match is refined for individual installations by looking at companies' internet websites (this was the case for 3 ETS installations). In addition, the following country-specific adjustments and assumptions are made:

- Data on N_2O *emissions* are retrieved from the Swiss Pollutant registry (which lists pollutants by installation), and subsequently deducted from verified ETS emissions. Only one installation under the ETS declared N_2O emissions in the Swiss Pollutant registry.
- *Process emissions* as declared by Switzerland to the UNFCCC are deducted from the industry sectors with process emissions as indicated in the Swiss 2012 National Inventory Report.
- The simple average of the two allowance auctions of 2014 is applied as the permit *price*. This method is chosen since no transactions of emissions certificates have yet taken place on the allowance trading platform.

- Since data on the Swiss Emissions Trading Scheme was taken for 2013, while the base data on emissions is taken for 2012 from the IEA Energy Balances, the resulting estimates might overestimate the emissions coverage of the Swiss ETS.
- The average emission permit *price* in 2013 amounts to CHF 30.20 per tonne CO_2. This translates into EUR 24.97 per tonne CO_2 when adjusted to the price level of 2012 using data from OECD (2010) and exchange rates from the TEU database.

B.3.8. USA

To determine the price signal sent by emissions trading systems in the USA an average auction price from RGGI and the California system was calculated for all sectors, weighted by the proportion of emissions subject to each system in that sector.

The weighted average auction price in the USA, expressed in 2012 prices, amounts to USD 12.35 (EUR 9.47) in the industry sector, USD 7.07 (EUR 5.42) in electricity and USD 12.92 (EUR 9.91) in the other sectors.

B.3.8.1. RGGI

Data on verified emissions of facilities is obtained from RGGI reports on annual emissions (RGGI, 2015b). In addition, the following system-specific adjustments and assumptions are made:

- All *emissions* are *attributed to the main industrial sector* of the facility as specified by the North American Industry Classification System (NAICS). The NAICS codes for facilities is obtained from the Energy Information Agency (EIA, 2015).
- The average auction clearing *price* in 2012 was USD 1.93 per tonne CO_2 (RGGI, 2015a), i.e. EUR 1.48 using exchange rates from the TEU database.

B.3.8.2. California

To estimate emissions subject to the California Cap-and-Trade Program in 2013, data from the Californian Air Resources Board (ARB) on the compliance obligations (in greenhouse gas emissions) of each entity that is subject to the programme is used (California EPA – ARB, 2014b). In addition, the following ETS-specific adjustments and assumptions are made:

- The *compliance obligations* of entities with more than one facility are split up into compliance obligations by facility according to the share of each facility's emissions in the entity's overall emissions. Emissions of each facility with an annual reporting obligation under the Mandatory Reporting of Greenhouse Gas Emissions Regulation can be found in California EPA – ARB (2014a). Note that all facilities subject to the Cap-and-Trade Program also have an annual reporting obligation.
- The *matching* of facilities to the six sectors discussed in the main body of the report is based on the North American Industry Classification System (NAICS). NAICS codes are reported for all facilities with an annual reporting obligation in California EPA – ARB (2014a).
- *Non-*CO_2 *and industrial process emissions* are excluded based on information from the California Greenhouse Gas Emission Inventory (California EPA – ARB, 2015c).

- As of 2015, fuel suppliers are included in California's Cap-and-Trade Program, which has a significant effect on the *coverage* of the road, and the offroad transport, residential and commercial, and agriculture and fisheries sectors. The extension of coverage to these sectors is included in the estimate.

  To approximate the extension of coverage, California's share in US total CO_2 emissions from energy use in the year 2012 is calculated for each of the four sectors and added to the estimates. Depending on data availabilities different datasets were used with a preference given to US national CO_2 emission balances as reported in EIA (2012).

  This implies that the extension of coverage in 2015 is estimated using 2012 data, which basically assumes a stable relationship over time between the size of CO_2 emissions in the respective sectors in California and the US. It further assumes that supply and demand could have resulted in the same permit price, had the coverage been extended in 2012 already. Finally, the approximation assumes a cap-and-trade coverage of 100% in these four sectors, an assumption confirmed by expert opinion from ARB (California EPA – ARB, 2015d).

- The average auction *price*[3] in 2013 was USD 12.83 per tonne CO_2 (California EPA – ARB, 2015), which translates into EUR 9.84 when adjusted to the price level of 2012 using data from OECD (2010) and exchange rates from the TEU database.

Notes

1. Exchange rates are from the OECD.Stat database (OECD, 2015).
2. Sector definitions used in this report and outlined in Section A.1.3 do not always match sector categories applied by individual countries.
3. Additional climate policies implemented in California, such as fuel economy standards, renewable portfolio standards, low-carbon fuel standards, are likely to increase allowance price volatility because they steepen the supply curve of abatement (Borenstein et al., 2015).

References and further sources

Borenstein, S. et al. (2015), "Expecting the Unexpected: Emissions Uncertainty and Environmental Market Design", *NBER Working Paper Series*, No. 20999, www.nber.org/papers/w20999.

California EPA – ARB (2015a), "Annual Allocation to Electrical Distribution Utilities (EDU) under the Cap-and-Trade Regulation", www.arb.ca.gov/cc/capandtrade/allowanceallocation/edu-ng-allowancedistribution/electricity-allocation.pdf (accessed 28 July 2015).

California EPA – ARB (2015b), "California Cap-and-Trade Program Summary of Auction Settlement Prices and Results", www.arb.ca.gov/cc/capandtrade/auction/results_summary.pdf (accessed 28 July 2015).

California EPA – ARB (2015c), *California Greenhouse Gas Emission Inventory - 2015 Edition* (database), www.arb.ca.gov/cc/inventory/data/tables/ghg_inventory_by_ipcc_00-13_20150424.xlsx (accessed 28 July 2015).

California EPA – ARB (2015d), Email communication with the Californian Air Resources Board, 6 October 2015.

California EPA – ARB (2014a), *Annual Summary of 2013 Greenhouse Gas Emissions Data Reported to the California Air Resources Board* (database), http://www.arb.ca.gov/cc/reporting/ghg-rep/reported-data/2013-ghg-emissions-2014-11-14-archive.xlsx (accessed on 28 July 2015).

California EPA – ARB (2014b), *2013 Compliance Obligation Detail for ARB's Cap-and-Trade Program* (database), www.arb.ca.gov/cc/capandtrade/2013compliancereport.xlsx (accessed on 28 July 2015).

California EPA – ARB (2013), "Vintage 2013, Industry Allowance Allocation by Sector", www.arb.ca.gov/cc/capandtrade/allowanceallocation/2013sectorallocation.pdf (accessed 28 July 2015).

California State Government, "California Cap on Greenhouse Gas Emissions and Market-Based Compliance Mechanisms", in *California Code of Regulations*, Title 17, Div. 3, Chap. 1, Sub chap. 10, Art. 5.

EIA (2015), Form EIA-860 detailed data for 2012, www.eia.gov/electricity/data/eia860/ (accessed 31 August 2015).

EIA (2012), State CO_2 Emissions by Energy Sectors, www.eia.gov/environment/emissions/state/ (accessed on 30 September 2015).

Environment Canada (2015), "Part 3", in *National Inventory Report: Greenhouse Gas Sources and Sinks in Canada 1990-2013*, Environment Canada, Ottowa.

EASA (2009), *SAVE - Study on Aviation and Economic Modelling - Final Report*, European Aviation Safety Agency, Cologne.

European Commission (2014a), Classification of installations in the EUTL Registry based on NACE 4 statistical classification, http://ec.europa.eu/clima/policies/ets/cap/leakage/studies_en.htm (accessed 5 April 2015).

European Commission (2014b), Verified Emissions for 2013, http://ec.europa.eu/clima/policies/ets/registry/documentation_en.htm (accessed 5 April 2015).

European Commission (2003), "Directive 2003/87/EC", *Official Journal of the European Union*, L275/32, European Commission, Brussels, Belgium.

European Energy Exchange (2015), Emission spot primary auction report 2012, https://www.eex.com/en/market-data/emission-allowances/auction-market/european-emission-allowances-auction/european-emission-allowances-auction-download (accessed 9 July 2015).

Federal Council of Switzerland (2012), Verordnung über die Reduktion der CO_2 Emissionen (CO_2 Ordinance), https://www.admin.ch/opc/de/classified-compilation/20120090/index.html (accessed 22 June 2015).

Federal Reserve Bank of St. Louis (2015), "Consumer Price Index, All items for China, CHNCPIALLMINMEI", in *FRED Economic Data* (database), https://research.stlouisfed.org/fred2/series/CHNCPIALLMINMEI/downloaddata (accessed 6 October 2015).

Gouvernement du Québec (2014), "Auction of Québec Greenhouse Gas Emission Units on December 3, 2013 - Summary Report Results", www.mddelcc.gouv.qc.ca/changements/carbone/ventes-encheres/resultats-vente20131203-en.pdf (accessed 13 July 2015).

Gouvernement du Québec (2012), "Determination of annual caps on greenhouse gas emission units relating to the cap-and-trade system for greenhouse gas emission allowances for the 2013-2020 period, Decret O.C. 1185-2012", *Gazette Officielle du Québec, Vol. 144, No. 51, p. 3612.*

Gouvernement du Québec (2013), Quantité d'unités d'émission versée en allocation gratuite en 2013 et liste des émetteurs qui en ont bénéficié, *Gazette Officielle du Québec*, Vol. 145, No. 48, pp. 1304-1305.

Gouvernement du Québec (2015), "Regulation respecting a cap-and-trade system for greenhouse gas emission allowances, chapter Q-2, s. 31, 1st par., subpars. b, c, d, e.1, h and h.1, ss. 46.1, 46.5, 46.6, 46.8 to 46.16, 115.27 and 115.34", in *Environment Quality Act*, www2.publicationsduquebec.gouv.qc.ca/dynamicSearch/telecharge.php?type=3&file=/Q_2/Q2R46_1_A.HTM (accessed 13 July 2015).

ICAP (2015), *Emissions trading Worldwide: Status Report 2015*, International Carbon Action Partnership, Berlin.

IPCC (2006), IPCC Guidelines for National Greenhouse Gas Inventories, Vol. 2 Energy, www.ipcc-nggip.iges.or.jp/public/2006gl (accessed 6 July 2015).

Korean Government (2012), Act on the Allocation and Trading of Greenhouse-Gas Emission Permits (English translation), http://elaw.klri.re.kr/kor_service/lawView.do?hseq=24561&lang=ENG (accessed 6 July 2015).

Korean Ministry of Environment (2014), National Emissions Allocation Plan 2015-2017, www.gir.go.kr/home/board/read.do?pagerOffset=10&maxPageItems=10&maxIndexPages=10&searchKey=&searchValue=&menuId=19&boardId=49&boardMasterId=8&boardCategoryId (accessed 6 July 2015).

Korean Exchange KRX (2015), KRU trading price, http://ets.krx.co.kr/contents/05/0502/JHPETS0502M01.jsp (accessed 6 July 2015).

Li, J. et al. (2015), "Integrating Cap and Trade into Taxing Energy Use: Interim Report for China", report prepared for the OECD.

Li, J. (2015), Personal Communication, 6 October, 2015.

National Bureau of Statistics of China (2014), "Energy Balances 2013", in *China Statistical Yearbook - 2014*, www.stats.gov.cn/tjsj/ndsj/2013/indexeh.htm.

NDRC (2011), Work Notice on Implementation of Carbon Emissions trading Pilot, Oct 29th 2011, www.sdpc.gov.cn/zcfb/zcfbtz/201201/t20120113_456506.html.

New Zealand Emission Unit Register (2015), Emissions Unit Register, https://app.eur.govt.nz/eats/nz/index.cfm?fuseaction=search.home (accessed 20 July 2015).

New Zealand Environmental Protection Authority (2015), Historical Carbon Market Information, www.epa.govt.nz/e-m-t/reports/carbon_market/Pages/historical_info.aspx (accessed on 2 October 2015).

New Zealand Ministry for the Environment (2015), New Zealand Emissions Trading Scheme, www.mfe.govt.nz/climate-change/reducing-greenhouse-gas-emissions/nz-emissions-trading-scheme (accessed 20 July 2015).

New Zealand Ministry for the Environment (2013), New Zealand's Greenhouse Gas Inventory, www.mfe.govt.nz/climate-change/reporting-greenhouse-gas-emissions/nzs-greenhouse-gas-inventory (accessed 20 July 2015).

New Zealand Ministry for the Environment (2012), "ETS 2012 Facts and Figures", *Info 695*, New Zealand Ministry for the Environment, Wellington.

New Zealand Ministry for the Environment (2009), "Climate Change (Stationary Energy and industrial Processes) Regulations 2009", *Emissions Trading Bulletin*, No. 13, www.mfe.govt.nz/sites/default/files/emissions-trading-bulletin-13.pdf.

New Zealand Government (2009), Climate Change (Stationary Energy and Industrial) Regulations 2009, http://legislation.govt.nz/regulation/public/2009/0285/latest/DLM2394207.html?search=ts_act%40bill%40regulation%40deemedreg_climate+change_resel_25_a&p=1 (accessed on 23 September 2015).

New Zealand Government (2008), Climate Change (Liquid Fossil Fuels) Regulations 2008, www.legislation.govt.nz/regulation/public/2008/0356/latest/DLM1635601.html (accessed on 23 September 2015).

OECD (2015), "Exchange rates", *OECD Stat* (database), http://stats.oecd.org (accessed 6 October 2015).

OECD (2010), "Main Economic Indicators - complete database" *Main Economic Indicators* (database), http://dx.doi.org/10.1787/data-00052-en (accessed 28 September 2015).

Ou, X., Zhang, X. and S. Chang, (2010), "Alternative fuel buses currently in use in China: Life-cycle fossil energy use, GHG emissions and policy recommendations", *Energy Policy*, Vol. 38, No. 1, pp. 406-418.

OTC-X (2015), CO_2 Emissionsrecht CHU (Allowance Trading Platform), https://www.otc-x.ch/markt/instrument/valor/999999.html (accessed 25 June 2015).

RGGI (2008), *Regional greenhouse gas initiative model rule: 12/31/08 with corrections*, Regional Greenhouse Gas Initiative, New York.

RGGI (2012), *Fact Sheet: RGGI CO_2 Allowance Auctions*, Regional Greenhouse Gas Initiative, New York.

RGGI (2015a), Allowance Prices and Volumes, https://www.rggi.org/market/co2_auctions/results (accessed 31 August 2015).

RGGI (2015b), Reports: Annual Emissions – Facility Level View, https://rggi-coats.org/eats/rggi/index.cfm?hc=ISkwICAK (accessed 27 August 2015).

RGGI (2015c), "Second Control Period CO2 Allowance Allocation", https://www.rggi.org/docs/CO2AuctionsTrackingOffsets/Allocation/SCP_Allowance-Allocation.pdf (accessed 19 October 2015).

SPG (2014a), Saitama 2012 Emission Data, www.pref.saitama.lg.jp/a0502/sakugen.html (accessed 20 August 2015).

SPG (2014b), Saitama 2012 Greenhouse Gas Inventory, www.pref.saitama.lg.jp/a0502/ontaih16co2.html (accessed 20 August 2015).

Swiss Emissions Trading Registry (2015), Public information, https://www.emissionsregistry.admin.ch/crweb/public/information.do (accessed 22 June 2015).

Swiss Federal Office for the Environment (2015a), Emissions trading scheme (ETS): information for ETS companies, www.bafu.admin.ch/klima/13877/14510/14719/14742/index.html?lang=de (accessed 22 June 2015).

Swiss Federal Office for the Environment (2015b), Email communication with the Swiss Federal Office for the Environment, 18 June 2015.

Swiss Federal Office for the Environment (2015c), Swiss Pollutant Release and Transfer Register, www.prtr.admin.ch/PublicWebSite/Abfrage.aspx?Lang=ger (accessed 25 June 2015).

TMG (2015), *Final Energy Consumption and Greenhouse Gas Emissions in Tokyo: FY 2012,* Bureau of Environment, Tokyo Metropolitan Government, Tokyo.

TMG (2014), "Tokyo Cap-and-Trade Program achieves 22% reduction after 3rd year", Bureau of Environment, Tokyo Metropolitan Government, Tokyo, www.kankyo.metro.tokyo.jp/en/climate/attachement/Tokyo%20C%26T%203rd%20Year%20Results.pdf (accessed 24 August 2015).

TMG (2012a), "Tokyo Cap-and-Trade Program for large facilities: Detailed Documents", Bureau of Environment, Tokyo Metropolitan Government, Tokyo, https://www.kankyo.metro.tokyo.jp/en/climate/attachement/Tokyo_Cap-and-Trade_Program_detailed_version.pdf (accessed 24 August 2015).

TMG (2012b), "Total reduction obligations and emissions trading scheme: Assessment of transaction prices", Bureau of Environment, Tokyo Metropolitan Government, Tokyo, www.kankyo.metro.tokyo.jp/climate/large_scale/attachement/trade_seminar121114_3_1.pdf (accessed 24 August 2015).

TMG (2012c), "The Tokyo Cap-and-Trade Program: Results for first fiscal year of operation", Bureau of Environment, Tokyo Metropolitan Government, Tokyo, https://www.kankyo.metro.tokyo.jp/en/climate/attachement/Result%20of%20the%20First%20FY%20of%20the%20Tokyo%20CT%20Program_final.pdf (accessed 20 August 2015).

TMG (2010), *Tokyo Cap-and-Trade Program: Japan's first mandatory emissions trading scheme,* Bureau of Environment, Tokyo Metropolitan Government, Tokyo.

UNFCCC (2015), "Submitted National Communications from non-Annex I Parties: Second National Communication on Climate Change of The People's Republic of China", http://unfccc.int/resource/docs/natc/chnnc2e.pdf (accessed 6 October 2015).

UNFCCC (2014), National Inventory Submissions 2014, http://unfccc.int/national_reports/annex_i_ghg_inventories/national_inventories_submissions/items/8108.php (accessed 6 July 2015).

World Bank (2013), "Tokyo's Emissions trading System. A review of its operation since 2000", in *Directions in Urban Development*, Urban Development and Resilience Unit, The World Bank, Washington, DC.

ORGANISATION FOR ECONOMIC CO-OPERATION AND DEVELOPMENT

The OECD is a unique forum where governments work together to address the economic, social and environmental challenges of globalisation. The OECD is also at the forefront of efforts to understand and to help governments respond to new developments and concerns, such as corporate governance, the information economy and the challenges of an ageing population. The Organisation provides a setting where governments can compare policy experiences, seek answers to common problems, identify good practice and work to co-ordinate domestic and international policies.

The OECD member countries are: Australia, Austria, Belgium, Canada, Chile, the Czech Republic, Denmark, Estonia, Finland, France, Germany, Greece, Hungary, Iceland, Ireland, Israel, Italy, Japan, Korea, Latvia, Luxembourg, Mexico, the Netherlands, New Zealand, Norway, Poland, Portugal, the Slovak Republic, Slovenia, Spain, Sweden, Switzerland, Turkey, the United Kingdom and the United States. The European Union takes part in the work of the OECD.

OECD Publishing disseminates widely the results of the Organisation's statistics gathering and research on economic, social and environmental issues, as well as the conventions, guidelines and standards agreed by its members.

OECD PUBLISHING, 2, rue André-Pascal, 75775 PARIS CEDEX 16
(23 2016 30 1 P) ISBN 978-92-64-26009-2 – 2016

www.ingramcontent.com/pod-product-compliance
Lightning Source LLC
Chambersburg PA
CBHW061214240926
51846CB00006B/1228
* 9 7 8 9 2 6 4 2 6 0 0 9 2 *